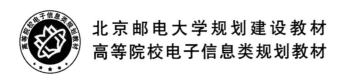

北京邮电大学规划建设教材
高等院校电子信息类规划教材

电路分析基础

（中英文版）

主　编　张雪菲
副主编　许晓东　范绍帅

北京邮电大学出版社
www.buptpress.com

内容简介

本书是匹配北京邮电大学通信工程留学生"电路分析基础"课程教学大纲的新型中英文双语教材。本书一共分6章,有中、英文两个版本。第1章电路元件与电路变量(重点内容为理想基本电路元件与分类、电路变量);第2章电路基本分析方法(重点内容为基尔霍夫电流定律、基尔霍夫电压定律、电阻的串并联组合、$2b$方法、节点电压法、网孔电流法);第3章电路基本理论(重点内容为叠加定理、电压源与电流源的等效变换、戴维南定理和诺顿定理、最大功率传输定理);第4章 RC 电路和 RL 电路(重点内容为动态元件特性、零输入响应与零状态响应、三要素法);第5章正弦稳态电路分析(重点内容为电路相量模型、基尔霍夫定律的相量形式、节点电压法和网孔电流法的相量形式、戴维南定理和诺顿定理的相量形式);第6章实验(重点内容为基本电路测量仪器介绍、实验实例)。

图书在版编目(CIP)数据

电路分析基础:中英文版 / 张雪菲主编. -- 北京:北京邮电大学出版社,2020.10
ISBN 978-7-5635-6189-6

Ⅰ. ①电… Ⅱ. ①张… Ⅲ. ①电路分析—高等学校—教材—汉、英 Ⅳ. ①TM133

中国版本图书馆 CIP 数据核字(2020)第 149975 号

策划编辑:刘纳新 姚 顺 责任编辑:刘 颖 封面设计:七星博纳

出版发行:北京邮电大学出版社
社　　址:北京市海淀区西土城路 10 号
邮政编码:100876
发 行 部:电话:010-62282185 传真:010-62283578
E-mail:publish@bupt.edu.cn
经　　销:各地新华书店
印　　刷:北京玺诚印务有限公司
开　　本:787 mm×1 092 mm　1/16
印　　张:13.75
字　　数:339 千字
版　　次:2020 年 10 月第 1 版
印　　次:2020 年 10 月第 1 次印刷

ISBN 978-7-5635-6189-6　　　　　　　　　　　　　　　　　定价:38.00 元
·如有印装质量问题,请与北京邮电大学出版社发行部联系·

本书主要面向初次接触电路知识的低年级本科生,旨在对电路课程中最基本的概念和分析方法进行讲解。本书为中、英文双语对照版本,适用于有出国留学打算的相关专业的学生和在华学习的留学生。通过对本书的学习,学生可以系统地学习基本的电子元件、电路分析原理和分析方法,特别是电路理论的一些基本定理和技术。此外,这本书还提供了4个与本书介绍知识相关的基础实验。希望通过对本书的学习,学生能够建立科学思维的意识和初步的工程评价能力,为后续电路课程(如电子电路基础和数字电路设计等)的学习打下良好的基础。

本书分为6章,分别从电路基本概念、电路基本分析方法、电路基本定理,以及不同类型元件、不同类型信号对电路性能的影响等方面进行了介绍。本书的内容为基础的电路知识,通过不断总结和横纵向对比,帮助读者更好地理解和学习较为抽象的电路知识。下面总结了本书所有章节的学习目标,帮助读者抓住学习的重点。

1. 第1章的学习目标:电路基本概念

(1) 了解电荷、电压、电流、功率与能量的定义以及它们之间的关系;

(2) 了解关联参考方向,并熟练掌握功率计算;

(3) 了解电阻、电容、电感、独立电源以及受控源的基本概念;

(4) 熟练掌握电阻、电容、电感的伏安特性关系(VCR)。

2. 第2章的学习目标:电路基本分析方法

(1) 理解节点、路径、回路、支路和网孔的定义;

(2) 理解电阻的串并联组合和等效方法;

(3) 熟练掌握基尔霍夫电流定律(KCL)和基尔霍夫电压定律(KVL);

(4) 理解和掌握2b方法;

(5) 熟练掌握节点电压法;

(6) 熟练掌握网孔电流法。

3. 第3章的学习目标:电路基本定理

(1) 理解并掌握叠加定理;

(2) 理解并掌握电压源电流源的等效变换;

(3) 理解并掌握戴维南定理和诺顿定理;

(4) 理解并掌握最大功率传输定理。

4. 第4章的学习目标:一阶动态电路

(1) 熟练掌握电感和电容的特性、VCR和串并联等效;

(2) 理解换路定则;

(3) 理解一阶电路、零输入响应电路和零状态响应电路的定义；
(4) 熟练掌握 RL 和 RC 电路的零输入响应的分析方法；
(5) 熟练掌握 RL 和 RC 电路的零状态响应的分析方法；
(6) 理解瞬时响应和稳态响应，熟练掌握完全响应的计算方法和三要素法。

5. 第 5 章的学习目标：正弦稳态电路

(1) 理解正弦信号、正弦电源和正弦响应的基本概念；
(2) 掌握正弦变量的相量表示方法；
(3) 理解频域内阻抗和导纳的概念；
(4) 理解并掌握频域上的电路基本分析方法和电路基本定理；
(5) 理解复数功率，并掌握复数功率计算方法。

6. 第 6 章的学习目标：针对本书前述知识点，通过软件或硬件实验来验证在前面几章学到的电路分析方法和电路定律

(1) 学习使用基本的电路测量仪器；
(2) 学习使用软件进行电路仿真；
(3) 验证电路分析方法和电路定律。

下面，请大家思考一下，究竟电路分析需要达到什么目的，也就是为什么要进行电路分析？在研究电路理论时，通常把它分为两类，即电路分析和电路设计。如图 1 所示，电路分析与电路设计之间是有明显差异的。电路分析是指已知输入信号和电路拓扑结构和参数的情况下，求解电路中某些输出响应的过程；而电路设计是指已知输入信号和输出信号，设计电路拓扑结构和参数的过程。尽管电路分析的目标集中在定量地分析电路变量，但同时它仍可以为电路设计的改进提供建议。在本书中，主要关注电路分析中涉及的知识介绍和分析方法的讲解。

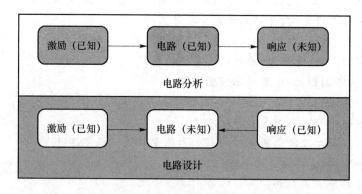

图 1 电路理论的范畴

电路分析是以数学理论为基础，预测电路模型及其理想电路元件的技术，它在电路设计中起着非常重要的作用。具体来说，这个技术是用数学模型为给定的电路建立方程，然后求解方程来确定电路中所需的电流和电压等。在建立方程时，主要根据拓扑约束和元件约束写出电路方程式。拓扑约束表示电路的拓扑关系(即连接形式)，元件约束描述电路中元件的特性。

本书由北京邮电大学张雪菲主编。由于编者水平有限，书中难免存在错误和不足之处，敬请读者批评指正，编者邮箱为 zhangxuefei@bupt.edu.cn。在本书的编写过程中，北京邮电大学的刘宝玲教授和吴建伟高工提供了宝贵的建议，在此谨致衷心的感谢！

目 录

第1章 电路元件和电路变量 ·· 1

  1.1 常见的电路变量 ·· 1

    1.1.1 电荷 ·· 1

    1.1.2 电压 ·· 1

    1.1.3 电流 ·· 2

    1.1.4 功率与能量 ·· 2

  1.2 基本电路元件 ·· 4

    1.2.1 电阻 ·· 4

    1.2.2 电容 ·· 5

    1.2.3 电感 ·· 6

    1.2.4 独立电压源与独立电流源 ·· 8

    1.2.5 受控源 ·· 8

  1.3 习题 ·· 9

第2章 电路基本分析方法 ·· 12

  2.1 节点、支路和网孔 ·· 12

  2.2 基尔霍夫电路定律 ·· 14

  2.3 电阻的串并联组合 ·· 16

  2.4 2b方法 ·· 20

  2.5 节点电压法 ·· 21

  2.6 网孔电流法 ·· 24

  2.7 习题 ·· 27

第3章 电路基本理论 ·· 33

  3.1 叠加定理 ·· 33

3.2　电压源与电流源的等效变换 ·· 35

3.3　戴维南定理和诺顿定理 ·· 38

3.4　最大功率传输定理 ·· 41

3.5　习题 ·· 42

第 4 章　RC 电路和 RL 电路　48

4.1　电感 ·· 48

4.2　电容 ·· 50

4.3　换路定则 ·· 51

4.4　一阶电路 ·· 52

4.5　RL 和 RC 电路的零输入响应 ·· 53

    4.5.1　RC 电路的零输入响应 ··· 53

    4.5.2　RL 电路的零输入响应 ··· 54

4.6　RL 和 RC 电路的零状态响应 ·· 56

    4.6.1　RL 电路的零状态响应 ··· 56

    4.6.2　RC 电路的零状态响应 ··· 57

4.7　RL 电路和 RC 电路的全响应和三要素法 ··· 58

4.8　习题 ·· 60

第 5 章　正弦稳态电路分析　66

5.1　正弦函数 ·· 66

5.2　正弦响应 ·· 67

5.3　相量 ·· 68

    5.3.1　相量变换 ··· 68

    5.3.2　相量反变换 ·· 68

    5.3.3　相量的基本运算 ·· 69

5.4　阻抗和导纳 ··· 70

5.5　电路元件 VCR 的相量表示 ··· 71

    5.5.1　电阻 VCR 的相量形式 ·· 71

    5.5.2　电容 VCR 的相量形式 ·· 71

    5.5.3　电感 VCR 的相量形式 ·· 72

5.6　基尔霍夫定律的相量形式 ·· 72

5.6.1 KCL 的相量形式 ………………………………………………………… 72

5.6.2 KVL 的相量形式 ………………………………………………………… 73

5.7 电路简化方法的相量形式 ……………………………………………………… 73

5.7.1 串联阻抗和并联阻抗 …………………………………………………… 73

5.7.2 电压源电流源的等效变换、戴维南定理和诺顿定理的相量形式 ……… 74

5.8 电路基本分析方法的相量形式 ………………………………………………… 75

5.8.1 节点电压法的相量形式 ………………………………………………… 75

5.8.2 网孔电流法的相量形式 ………………………………………………… 76

5.9 功率计算 ………………………………………………………………………… 77

5.10 最大功率传输定理的相量形式 ………………………………………………… 78

5.11 习题 ……………………………………………………………………………… 79

第6章 实验 …………………………………………………………………………… 84

6.1 基本电路测量仪器介绍 ………………………………………………………… 84

6.1.1 万用表 …………………………………………………………………… 84

6.1.2 直流电源 ………………………………………………………………… 85

6.1.3 电阻和滑动变阻器 ……………………………………………………… 85

6.1.4 电容和电感 ……………………………………………………………… 86

6.1.5 示波器 …………………………………………………………………… 86

6.1.6 数字信号发生器 ………………………………………………………… 86

6.2 电路仿真软件介绍 ……………………………………………………………… 87

6.3 实验实例 ………………………………………………………………………… 87

6.3.1 电阻的直流特性 ………………………………………………………… 87

6.3.2 验证基尔霍夫定律 ……………………………………………………… 90

6.3.3 RC 电路和 RL 电路的测量 ……………………………………………… 92

6.3.4 验证戴维南定理的实验 ………………………………………………… 94

Chapter 1 Circuit Elements and Circuit Variables ……………………………… 101

1.1 Common Circuit Variables ……………………………………………………… 101

1.1.1 Electric charge …………………………………………………………… 101

1.1.2 Voltage …………………………………………………………………… 102

1.1.3 Current …………………………………………………………………… 102

1.1.4　Power and energy ……………………………………………………… 103

1.2　Basic circuit element ……………………………………………………… 105

    1.2.1　Resistor ……………………………………………………………… 105

    1.2.2　Capacitor ……………………………………………………………… 106

    1.2.3　Inductor ……………………………………………………………… 107

    1.2.4　Independent voltage source and independent current source ……… 109

    1.2.5　Dependent source …………………………………………………… 110

1.3　Exercises …………………………………………………………………… 111

Chapter 2　Basic Circuit Analysis Method ………………………………… 114

2.1　Node, branch and mesh …………………………………………………… 114

2.2　Kirchhoff's circuit law ……………………………………………………… 116

2.3　Series-parallel combinations of resistors ………………………………… 119

2.4　$2b$-method ………………………………………………………………… 123

2.5　Node voltage method ……………………………………………………… 125

2.6　Mesh current method ……………………………………………………… 128

2.7　Exercises …………………………………………………………………… 131

Chapter 3　Circuit Theorem …………………………………………………… 137

3.1　Superposition theorem ……………………………………………………… 137

3.2　Source transformations …………………………………………………… 139

3.3　Thevenin theorem and Norton theorem ………………………………… 143

3.4　Maximum power transfer theorem ………………………………………… 145

3.5　Exercises …………………………………………………………………… 147

Chapter 4　*RC* Circuit and *RL* Circuit ……………………………………… 153

4.1　First-order circuits ………………………………………………………… 153

4.2　Capacitor …………………………………………………………………… 155

4.3　Switching rule ……………………………………………………………… 156

4.4　First-order circuit ………………………………………………………… 157

4.5　Zero-input response of *RL* and *RC* circuit ……………………………… 158

    4.5.1　Zero-input response of *RC* circuit ………………………………… 158

  4.5.2 Zero-input response of RL circuit ········· 160
4.6 Zero-state response of RL and RC circuit ········· 162
  4.6.1 Zero-state response of RL circuit ········· 162
  4.6.2 Zero-state response of RC circuit ········· 163
4.7 Complete response and three element method for RL and RC circuit ········· 164
4.8 Exercises ········· 167

Chapter 5 Sinusoidal Steady-State Analysis ········· 173

5.1 Sinusoidal function ········· 173
5.2 Sinusoidal response ········· 174
5.3 The phasor ········· 175
  5.3.1 Phasor transformation ········· 175
  5.3.2 Inverse phasor transformation ········· 176
  5.3.3 Basic operations of phasors ········· 176
5.4 Passive circuit elements in phasor ········· 177
5.5 VCR of circuit element in phasor ········· 179
  5.5.1 VCR of resistor in phasor ········· 179
  5.5.2 VCR of capacitor in phasor ········· 179
  5.5.3 VCR of inductor in phasor ········· 179
5.6 Kirchhoff's Law in phasor ········· 180
  5.6.1 KCL in phasor ········· 180
  5.6.2 KVL in phasor ········· 181
5.7 Circuit simplifications in phasor ········· 181
  5.7.1 Impedances in series and parallel ········· 181
  5.7.2 Source transformations in phasor, Thevenin theorem in phasor and Norton theorem ········· 182
5.8 Basic circuit analysis methods in phasor ········· 183
  5.8.1 Node voltage method in phasor ········· 183
  5.8.2 Mesh current method in phasor ········· 184
5.9 Power calculation ········· 185
5.10 Maximum power transfer theorem in phasor ········· 187
5.11 Exercises ········· 188

Chapter 6 Experiment ········ 193

6.1 Basic electronic measuring devices ········ 193
6.1.1 Multi-meter ········ 193
6.1.2 DC power supply ········ 194
6.1.3 Resistor and slide rheostat ········ 194
6.1.4 Capacitor and inductor ········ 195
6.1.5 Oscilloscope ········ 195
6.1.6 Digital signal generator ········ 195

6.2 The introduction of circuit simulation software ········ 196

6.3 Experiment instances ········ 196
6.3.1 DC characteristics of resistors ········ 196
6.3.2 verification for Kirchhoff's laws ········ 199
6.3.3 *RC* circuit and *RL* circuit ········ 201
6.3.4 Verification for Thevenin theorem ········ 204

第1章 电路元件和电路变量

电路分析的学习可以被描述为对不同电路元件根据不同拓扑结构组成的电路进行电路变量定量分析的过程。因此,本章将介绍电路理论中最基本的知识,即电路变量和电路元件。

第1章的学习目标:电路基本概念
(1) 了解电荷、电流、电压、功率和能量的定义以及它们之间的关系;
(2) 了解关联参考方向,并熟练掌握功率计算;
(3) 了解电阻、电容、电感、独立电源以及受控源的基本概念;
(4) 熟练掌握电阻、电容、电感的伏安特性关系(VCR)。

1.1 常见的电路变量

本节介绍了电路学习中最常见和最基本的电路变量,主要包括电荷、电压、电流、功率与能量,以及它们的计算方法、它们之间的转化关系等。

1.1.1 电荷

电荷的概念是描述所有电现象的基础。电荷是双极性的,根据电场作用力的方向性,电荷可分为正电荷与负电荷。在高中物理中,已经学习过正电荷和负电荷的划分,即规定用丝绸摩擦过的玻璃棒带的是正电荷,用毛皮摩擦过的橡胶棒带的是负电荷。

在物理学中,电荷的多少称为电荷量。在国际单位制里,电荷量的符号以 Q 为表示,单位是库仑(C)。库仑是单个电子所带的负电荷量($1.602\,189\,2 \times 10^{-19}$ 库仑)的整数倍,也就是说1库仑相当于 $6.241\,46 \times 10^{18}$ 个电子所带的电荷总量。

1.1.2 电压

电压也被称为电势差或电位差,是衡量单位电荷在静电场中由于电势不同所产生的能量差的物理量。其大小等于单位正电荷因受电场力作用从 A 点移动到 B 点所做的功,电压的方向规定为从高电位指向低电位的方向。电压的国际单位制为伏特(V,简称伏),常用的

单位还有毫伏(mV)、微伏(μV)、千伏(kV)等。电压的定义可被记作

$$v = \frac{\mathrm{d}w}{\mathrm{d}q} \tag{1.1}$$

其中，v 是以伏特(V)为单位表示的电压值，w 是以焦耳(J)为单位表示的能量值，而 q 是以库伦(C)为单位表示的电荷量。

如图 1.1 所示，A、B 两节点之间的电压差 v_{AB} 等于从 B 节点移动一个单位电荷到 A 节点所需要的能量。

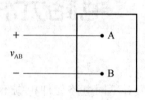

图 1.1　一个二端元件

任何电压的定义必须包括电压幅度值和代表极性(高电势和低电势)的正负号对。为了分析的简洁性，本书定义电压的参考极性是由某个正负号对指定的任意极性。

1.1.3　电流

电流也被称为电流强度，是衡量电荷流动的速度的物理量，其大小是单位时间里通过导体任一横截面的电量。电流的方向规定为正电荷(质子)的运动方向，但从宏观上很难知道正电荷的运动方向。所以，定义了电流的参考方向，它是由一个箭头指定的任意的方向。电流的国际单位制为安培(A，简称安)，常用的单位还有毫安(mA)和微安(μA)等。电流的定义可被记作

$$i = \frac{\mathrm{d}q}{\mathrm{d}t} \tag{1.2}$$

其中，i 是以安培(A)为单位表示的电流值，q 是以库伦(C)为单位表示的电荷量，而 t 是以秒(s)为单位表示的时间。

任何电流的定义必须包括电流幅度值和电流方向。

问题 1.1　假设穿过某电子元件的电流为 2 A，计算 1 min 内进入该元件的电荷量的大小。

解：1 min 内进入该元件的电荷量大小为：2 A×60 s＝120 C。

1.1.4　功率与能量

功率是单位时间内消耗或吸收的能量，可被记作

$$P = \frac{\mathrm{d}w}{\mathrm{d}t} \tag{1.3}$$

其中，P 是以瓦特(W)为单位表示的功率值，w 是以焦耳(J)为单位表示的能量值，而 t 是以秒(s)为单位表示的时间。

在进一步讨论之前,本书先介绍另外一个概念——**关联参考方向**,它对功率的计算很重要。在上述讨论电压和电流时,电压的参考极性和电流的参考方向都是可以任意设定的。但是一旦被规定,就必须在之后的分析中遵循这些规定。在这些符号约定之中,使用最广泛的是关联参考方向。

当电流的参考方向是沿着元件上电压参考极性的电势下降方向时,符合关联参考方向,则关于电压与电流关系的任何表达式中都使用正号;否则,使用负号。在本书的所有分析中都将遵循关联参考方向。

由电压和电流的定义,式(1.3)可重写为

$$P=\left(\frac{\mathrm{d}w}{\mathrm{d}q}\right)\left(\frac{\mathrm{d}q}{\mathrm{d}t}\right)=vi \tag{1.4}$$

请注意式(1.4)成立的前提是要遵循关联参考方向。如果不遵循,即电流的参考方向指向电压参考极性电势上升的方向,则功率表示为

$$P=-vi \tag{1.5}$$

当 $P>0$ 时,能量被该电子元件吸收,或者说能量被传递给电子元件;当 $P<0$ 时,该电子元件产生能量,或者说能量从该元件中释放出来。

能量是功率在时间上的累积,功率是单位时间内消耗或吸收的能量,其关系可表示为

$$w=\int_{-\infty}^{t}P\mathrm{d}t \tag{1.6.a}$$

或者

$$P=\frac{\mathrm{d}w}{\mathrm{d}t} \tag{1.6.b}$$

电路中基础变量的通用单位如表 1.1 所示。

表 1.1 电路中基础变量的通用单位

名 称	通用单位
电压	伏特(V),毫伏($mV=10^{-3}$ V),微伏($\mu V=10^{-6}$ V)
电流	安培(A),毫安($mA=10^{-3}$ A),微安($\mu A=10^{-6}$ A)
功率	瓦特(W=V·A),毫瓦($mW=10^{-3}$ W)

问题 1.2 假设黑匣子内有某个未知电路,现针对其中的两个节点(即节点 1 和节点 2)进行分析。已知节点 2 与节点 1 之间的电压为 20 V,同时流入节点 2 的电流值为 4 A。

求:(1) 根据图 1.2 所示的参考极性,分别计算 v 和 i 的值。

(2) 请说明黑匣子内的电路是吸收能量还是产生能量?

解:(1) 在子图(a)中,$v=-20$ V,$i=-4$ A;在子图(b)中,$v=-20$ V,$i=4$ A;在子图(c)中,$v=20$ V,$i=-4$ A;在子图(d)中,$v=20$ V,$i=4$ A。

(2) 在子图(a)中,$P=vi=(-20\text{ V})(-4\text{ A})=80\text{ W}>0$,吸收能量;在子图(b)中,$P=-vi=-(-20\text{ V})(4\text{ A})=80\text{ W}>0$,吸收能量;在子图(c)中,$P=-vi=-(20\text{ V})(-4\text{ A})=80\text{ W}>0$,吸收能量;在子图(d)中,$P=vi=20\text{ V}\cdot 4\text{ A}=80\text{ W}>0$,吸收能量。

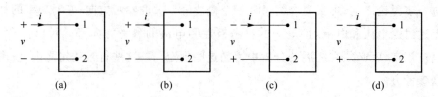

图 1.2　问题 1.2(1)的电路图

问题 1.3　分别计算图 1.3 中三个子图所示电路的功率。

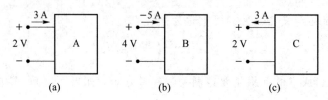

图 1.3　问题 1.3 中的电路图

解：(a) 服从关联参考方向，$P=(2\text{ V})(3\text{ A})=6\text{ W}$。由于功率是正值，说明能量被 A 吸收或者说被传递给 A。

(b) 服从关联参考方向，$P=(4\text{ V})(-5\text{ A})=-20\text{ W}$。由于功率是负值，说明 B 产生能量或者说能量从 B 中释放出来。

(c) 不服从关联参考方向，$P=-(2\text{ V})(3\text{ A})=-6\text{ W}$。由于功率是负值，说明 C 产生能量或者说能量从 C 中释放出来。

1.2　基本电路元件

本节将介绍电路中的重要组成部分之一——电路元件。一个理想的基本二端电路元件具备以下三个属性：

(1) 只有两个节点，且这两个节点是可以与其他电子元件相连的连接点；

(2) 电路分析中，是以电流和（或）电压来描述其外部电气特性，而忽略了一些其他特性，如热特性、电磁影响等；

图 1.4　一个理想的基本电路元件

(3) 最小元件单元，即它无法再被细分为其他的电路元件。

在这里，"基本"一词意味着这种电路元件无法被进一步简化或细分为其他电路元件；"理想"一词意味着这种电路元件在现实中往往是不存在的。但为了简化分析，在本书中所有的基本电路元件都可以被认为是理想的基本电路元件。图 1.4 是一个理想的基本电路元件的表示。

1.2.1　电阻

电阻是一种无源双端电路元件。在电路中，电阻主要用于减小电流、调整信号电压、分压，以及有源元件的偏压等。图 1.5 为电阻在电路中的符号和实际的电阻元件，在电路图中通常用字母 R 表示电阻元件。电阻的大小用电阻值表示。对于截面均匀的电阻，其电阻

值为

$$R=\rho\frac{L}{A} \tag{1.7}$$

其中，R 为以欧姆（Ω）为单位表示的电阻值，ρ 为电阻材料的电阻率（Ω·cm），L 为电阻体的长度（cm），A 为电阻体的截面积（cm²）。

图 1.5　电阻的符号和实际的电阻元件

电阻的电压电流关系也被称为欧姆定律，即电阻两端的电压等于它的电阻值乘以流过它的电流，且电压电流需要服从关联参考方向。具体可表示为

$$v=iR \tag{1.8}$$

电阻的功率可表示为

$$P=vi=i^2R=\frac{v^2}{R} \tag{1.9}$$

1.2.2　电容

电容是一种在电场中储存能量的无源双端电路元件，在调谐、旁路、耦合、滤波等电路中起着重要的作用。电容可用于构造动态数字存储器，制作高通或低通滤波器，或者用于通交流阻直流电流。图 1.6 为电容的符号及其实际的元件，在电路图中通常用字母 C 表示电容元件。

图 1.6　电容的符号及其实际的元件

当两个平行的相互靠近的金属极板，中间夹一层不导电的绝缘介质，就构成了电容器。当电容器的两个极板之间加上电压时，电容器就会储存电荷。电容器的电容量在数值上等于一个导电极板上的电荷量与两个极板之间的电压之比，具体可记为

$$C=\frac{Q}{v} \tag{1.10}$$

其中，C 为以法拉（F）为单位表示的电容值，Q 为电荷量〔单位为库伦（C）〕；v 为电压〔单位为伏（V）〕。

电容的电压电流关系为流过电容的电流等于它两端的电压的变化率乘以电容值，且电压电流需要服从关联参考方向。在此情况下，具体可表示为

$$i=C\frac{\mathrm{d}v}{\mathrm{d}t} \tag{1.11}$$

如果电压电流不服从关联参考方向,则式(1.11)需要添加负号。

问题 1.4 当 $t<0$ 时,容值为 $0.6\ \mu\text{F}$ 的电容两端的电压为零。当 $t\geqslant 0$ 时,该电容两端的电压为 $40\text{e}^{-15\,000t}\sin 30\,000t\ \text{V}$。

(1) 求 $i(0)$;

(2) 求 $t=\pi/80$ ms 时,电容的功率值;

(3) 求 $t=\pi/80$ ms 时,电容存储的能量值。

解:

(1) $i=C\dfrac{\text{d}v}{\text{d}t}=(0.72\cos 30\,000t-0.36\sin 30\,000t)\text{e}^{-15\,000t}\ \text{A}, i(0)=0.72\ \text{A}$;

(2) $i(\pi/80\ \text{ms})=-31.66\ \text{mA}, v(\pi/80\ \text{ms})=20.505\ \text{V}, p=vi=-649.23\ \text{mW}$;

(3) $w=\displaystyle\int_{-\infty}^{t} p\,\text{d}t=\dfrac{1}{2}Cv^2=126.13\ \mu\text{J}$。

1.2.3 电感

电感器是能够把电能转化为磁能并存储起来的元件,电感器是一个用导线绕成的线圈。它的结构类似于变压器,但只有一个绕组。图 1.7 给出了电感器的符号及其实际的元件,在电路图中通常用字母 L 表示电感元件,它的单位是亨利(H),常用的单位还有毫亨(mH)和微亨(μH)。

电感的电压电流关系为电感两端的电压等于"流过"它的电流的变化值乘以电感值,且电压电流需要服从关联参考方向。具体可表示为

$$v=L\dfrac{\text{d}i}{\text{d}t} \tag{1.12}$$

问题 1.5 如图 1.8 所示电路,其中电流源在 $t<0$ 时产生的电流为零,在 $t>0$ 时产生电流为 $10t\text{e}^{-5t}\ \text{A}$。

(1) 画出电流的波形图;

(2) 写出 100 mH 电感两端的电压;

(3) 画出电压的波形图。

图 1.7 电感器的符号及其实际的元件　　图 1.8 问题 1.5 中的电路

解:

(1) 电流波形如图 1.9 所示。

(2) 当 $t>0$ 时,$v=L\dfrac{\text{d}i}{\text{d}t}=\text{e}^{-5t}(1-5t)\ \text{V}$;当 $t<0$ 时,$v=0$。

(3) 电压的波形如图 1.10 所示。

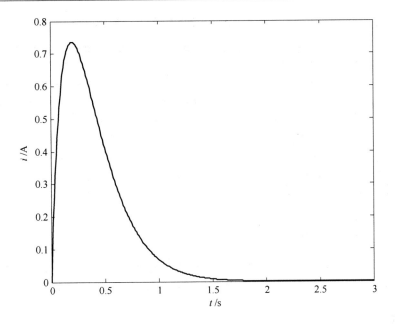

图 1.9 问题 1.5(a)的图形

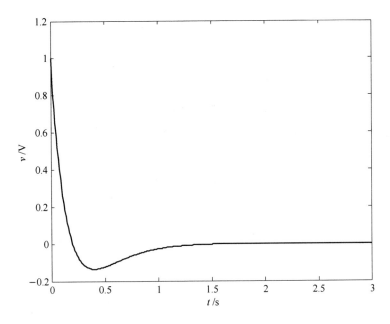

图 1.10 问题 1.5(c)的图形

基本电路元件的通用单位如表 1.2 所示。

表 1.2 基本电路元件的通用单位

元件名称	通用单位
电阻	欧姆(Ω),千欧($k\Omega$),兆欧($M\Omega$)
电容	法拉(F),皮法(pF),微法(μF)
电感	亨利(H),毫亨(mH),微亨(μH)

电阻、电容、电感的 VCR、功率及能耗的计算如表 1.3 所示。

表 1.3 电阻、电容、电感的 VCR、功率及能耗的计算(服从电压电流的关联参考方向)

元件名称	VCR	功率	能耗
电阻	$v=iR$(欧姆定律)	$P=vi=i^2R=\dfrac{v^2}{R}$	$w=Pt=i^2Rt=\dfrac{v^2}{R}t$
电容	$i=C\dfrac{dv}{dt}$	$P=vi$	$w=\displaystyle\int_{-\infty}^{t}Pdt=\dfrac{1}{2}Cv^2$
电感	$v=L\dfrac{di}{dt}$	$P=vi$	$w=\displaystyle\int_{-\infty}^{t}Pdt=\dfrac{1}{2}Li^2$

1.2.4 独立电压源与独立电流源

一个理想的独立电压源是一个不管其两端的电流如何流动,都能在两端之间维持一个固定的电压值的二端元件。图 1.11(a)和(b)分别为独立电压源的符号和输出特性。

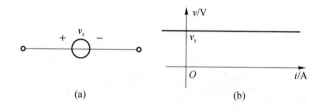

图 1.11 独立电压源的符号和输出特性

理想的独立电流源是一种不管其两端的电压如何变化,都能在其两端之间维持一个事先设定好的电流值的二端元件。图 1.12(a)和(b)分别为独立电流源的符号和输出特性。

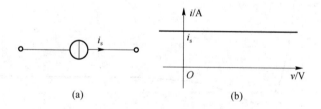

图 1.12 独立电流源的符号和输出特性

1.2.5 受控源

独立电源在电路中提供电压或电流是不依赖于电路中其他位置的电压或电流的,与其相反,受控源提供的电压或电流是由电路中其他位置的电压(或电流)决定。常见的受控源可分为四类:电压控制的电压源、电流控制的电压源、电压控制的电流源和电流控制的电流源,分别如图 1.13(a)、(b)、(c)和(d)所示。

一个电压控制的电压源所提供的电压由电压控制的系数、控制电压和产生电压之间的关系,以及产生电压的参考极性这三部分确定。在图 1.13(a)中,控制电压和产生电压 v_s 之

间的计算公式为
$$v_s = \mu v_x \tag{1.13}$$
其中,控制电压的符号为 v_x,μ 是电压比系数(一个无量纲的比例系数)。v_s 的参考极性如图 1.13(a)所示。

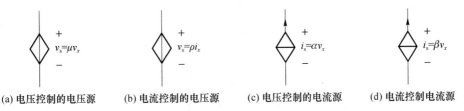

(a) 电压控制的电压源　　(b) 电流控制的电压源　　(c) 电压控制的电流源　　(d) 电流控制电流源

图 1.13　受控电源的符号

电压控制的电流源产生的电压 v_s 的计算公式为
$$v_s = \rho i_x \tag{1.14}$$
其中,i_x 表示控制电流的大小,ρ 是转移电阻(一个单位是 V/A 的系数)。

电压控制的电流源产生的电流 i_s 的计算公式为
$$i_s = \alpha v_x \tag{1.15}$$
其中,v_x 表示控制电压的大小,α 是转移电导(一个单位是 A/V 的系数)。

电流控制的电流源产生的电流 i_s 的计算公式为
$$i_s = \beta i_x \tag{1.16}$$
其中,i_x 表示控制电流的大小,β 是电流比系数(一个无量纲的比例系数)。

1.3　习　　题

习题 1.1　流入某元件的电流值为 $i(t) = 2\cos 2\,000t$ A,求 $q(t)$ 的表达式。

习题 1.2　如图 1.14 所示,连接黑匣 A 和黑匣 B 中的两个电路中,电流的参考方向和电压的参考极性如图所示。对于下列每组数值,分别计算系统的功率,并说明能量是从 A 流向 B 还是从 B 流向 A。

(1) $i = 2$ A,$v = 220$ V;

(2) $i = -3$ A,$v = 250$ V;

(3) $i = 5$ A,$v = -20$ V;

(4) $i = -5$ A,$v = -10$ V。

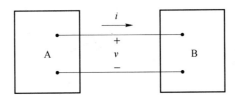

图 1.14　习题 1.2 中的电路

习题 1.3　在 $t < 0$ 时,电路元件两端的电压和电流为零;在 $t \geqslant 0$ 时,电压 $v = 5\mathrm{e}^{-50t}\sin 150t$ V,

电流 $i=2\mathrm{e}^{-50t}\sin 150t$ A。求该元件在 $t=10$ ms 时吸收的能量值。

习题 1.4　如图 1.15 所示,电流 $i_s=2$ A,计算受控源的电压值。

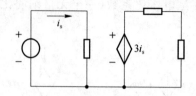

图 1.15　习题 1.4 中的电路

习题 1.5　假设当 $0<t<5\pi$ ms 时,某一 20 μF 电容的两端电压为 $v=50\sin 200t$ V,求电荷量、功率和能量,并绘制能量的波形图。

习题 1.6　如图 1.16 所示,元件 A 的电压、电流分别为 $v=-3$ V、$i=5$ A,元件 B 的电压、电流分别为 $v=5$ V 和 $i=-2$ mA,求两元件吸收的功率大小。

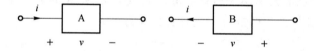

图 1.16　习题 1.6 中的电路

习题 1.7　电路如题图 1.17 所示,其中 v_s 为理想电压源,试判断若外电路不变,仅电阻 R 变化时,将会引起电路中哪条支路电流的变化。

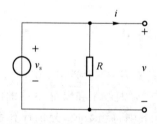

图 1.17　习题 1.7 中的电路

习题 1.8　一段含源支路及其电压电流特性如图 1.18 所示,图中 3 条直线对应于电阻 R 的 3 个不同数值 R_1、R_2 和 R_3,试根据该图判断各电阻的大小关系。

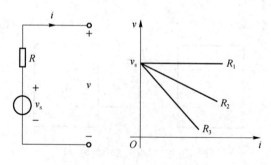

图 1.18　习题 1.8 中的电路

习题 1.9 如图 1.19 所示,若 $v_s>0, i_s>0, R>0$,请指出电路中各元件是吸收功率还是供出功率。

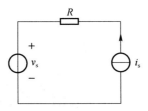

图 1.19 习题 1.9 中的电路

习题 1.10 如图 1.20 所示,试验证该电路功率守恒。

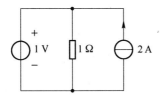

图 1.20 习题 1.10 中的电路

习题 1.11 图 1.21 所示的电路中,若电压源 $v_s=10\text{ V}$,电流源 $i_s=1\text{ A}$,试说明电压源和电流源是否一定都在提供功率。

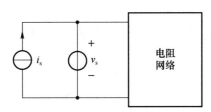

图 1.21 习题 1.11 中的电路

电路基本分析方法

序言中曾提到,电路分析主要是在电路的拓扑约束和元件约束下的分析,本章将介绍基于拓扑约束的电路基本分析方法。

第 2 章的学习目标:电路基本分析方法
(1) 理解节点、路径、回路、网孔和支路的定义;
(2) 理解串并联组合和等效方法;
(3) 熟练掌握基尔霍夫电流定律(KCL)和基尔霍夫电压定律(KVL);
(4) 理解和掌握 $2b$ 方法;
(5) 熟练掌握节点电压法;
(6) 熟练掌握网孔电流法。

2.1 节点、支路和网孔

在本书中,默认所有的电路都是平面电路,也就是那些可以在平面上绘制的没有交叉支路的电路,如图 2.1(a)所示。下面介绍分析平面电路时的一些重要的概念。

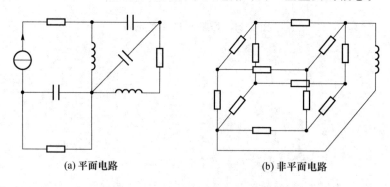

(a) 平面电路　　　　　　　(b) 非平面电路

图 2.1 平面电路和非平面电路

1. 节点

节点,是指两个或两个以上的电子元件公共连接的点。每个二端元件的两端都各有一个节点。值得注意的是,所有的理想导线都被视作某个节点的一部分。

如图 2.2 所示,根据节点的定义,点 A 和 H 为同一个节点,点 F、D 和 G 为同一个节点。因此,图 2.2 中有 5 个节点。

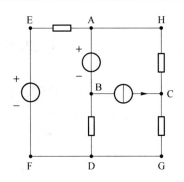

图 2.2　一个电路模型实例

2. 路径、回路和支路

从某一节点出发沿着一些支路连续移动,从而到达另一节点。这样一系列支路构成一条路径。

如果关于某条路径的集合中,开始的节点与末尾的节点相同,那么该路径就被称为**回路**。回路是一条封闭的路径。

支路是指由一个简单元件和该元件两端的节点组成的一条单一路径。根据支路的定义可知,支路的数量要等于元件的数量。因此,图 2.2 中有 7 个支路。

3. 网孔

网孔,是指其中不包含任何其他回路的一条回路。同时,网孔内部不包含任何元件。因此,图 2.3 中有 3 个网孔。

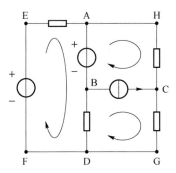

图 2.3　一个电路模型实例

分别找出图 2.4 中的支路、节点和网孔的个数。图 2.4（a）中有 14 个支路,9 个节点,6 个网孔;图 2.4（b）中有 10 个支路,7 个节点,4 个网孔。

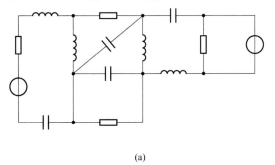

(a)

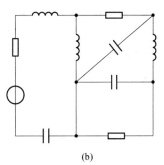

(b)

图 2.4　电路模型实例

2.2 基尔霍夫电路定律

电荷守恒和能量守恒是自然界的基本法则,它们在集总电路中的体现就是基尔霍夫电路定律。基尔霍夫电路定律适用于直流电路以及电路尺寸远小于电磁辐射的波长的交流电路。基尔霍夫电路定律分为基尔霍夫电压定律和基尔霍夫电流定律,分别代表了电路中电压和电流的约束关系。下面对于这两个定律进行详细的介绍。

1. 基尔霍夫电流定律

基尔霍夫电流定律,简称 KCL,具体可表示为:在任一时刻,电路的任一节点,进入该节点的电流的代数和为零,即 $\sum_{n=1}^{N} i_n(t) = 0$。或者说,流入该节点的电流之和等于流出该节点的电流之和。请注意,电流的参考方向对于 KCL 是至关重要。尽管电流的参考方向可以任意设定,但是在之后的分析中需要遵从设定的参考方向。KCL 实例如图 2.5 所示。

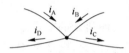

图 2.5 KCL 实例

根据 KCL,图 2.5 中的 KCL 方程可被表示为

$$i_A + i_B + (-i_C) + (-i_D) = 0 \quad (2.1.a)$$

或

$$(-i_A) + (-i_B) + i_C + i_D = 0 \quad (2.1.b)$$

或

$$i_A + i_B = i_C + i_D \quad (2.1.c)$$

问题 2.1 根据 KCL,计算图 2.6 所示电路中的电流值。

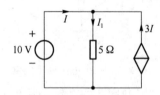

图 2.6 问题 2.1 中的电路

解:使用欧姆定律计算电流 I_1,

$$I_1 = \frac{10 \text{ V}}{5 \text{ }\Omega} = 2 \text{ A} \quad (2.2)$$

根据 KCL,

$$I_1 = I + 3I \quad (2.3)$$

得

$$I = 0.5 \text{ A} \quad (2.4)$$

2. 基尔霍夫电压定律

基尔霍夫电压定律,简称 KVL,具体可表示为:在任一时刻,电路中的任一回路中,各支路电压代数和为零,即 $\sum_{n=1}^{N} v_n(t) = 0$。请注意,在 KVL 中参考方向和参考极性都需考虑。KVL 中的参考方向可以是顺时针的也可以是逆时针的。

根据 KVL,图 2.7 中 a 和 b 两个参考方向下的 KVL 方程分别被表示为

$$-v_1 + v_2 - v_3 = 0 \qquad (2.5.\text{a})$$

或

$$v_1 + v_3 - v_2 = 0 \qquad (2.5.\text{b})$$

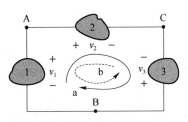

图 2.7 KVL 实例

问题 2.2 计算图 2.8 中电路的电流 I_0 并计算电路中各个元件的功率。

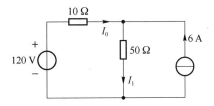

图 2.8 问题 2.2 中的电路

解:根据 KCL 和 KVL,可得 $I_1 - I_0 - 6\text{ A} = 0, 50\text{ }\Omega I_1 + 10\text{ }\Omega I_0 - 120\text{ V} = 0$。继而求得 $I_0 = -3\text{ A}, I_1 = 3\text{ A}$。

电流源的功率为

$$P_I = -v_1(6\text{ A}) = -150\text{ V}(6\text{ A}) = -900\text{ W} \qquad (2.6)$$

电压源的功率为

$$P_V = -120\text{ V}I_0 = 360\text{ W} \qquad (2.7)$$

阻值为 50 Ω 的电阻的功率为

$$P_{R=50\text{ }\Omega} = 50\text{ }\Omega(3\text{ A})^2 = 450\text{ W} \qquad (2.8)$$

阻值为 10 Ω 的电阻的功率为

$$P_{R=10\text{ }\Omega} = 10\text{ }\Omega(-3\text{ A})^2 = 90\text{ W} \qquad (2.9)$$

问题 2.3 计算图 2.9 所示电路中电压源、电流源和电阻的功率。

解:根据欧姆定律和 KVL,电流源的电压为

$$v_s = v_r - 20\text{ V} = 8\text{ V} - 20\text{ V} = -12\text{ V} \qquad (2.10)$$

电流源的功率为

$$P_I = -(-12\text{ V}) \times 2\text{ A} = 24\text{ W} \qquad (2.11)$$

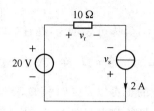

图 2.9 问题 2.3 中的电路

电压源的功率为

$$P_V = -20\text{ V} \times 2\text{ A} = -40\text{ W} \tag{2.12}$$

电阻的功率为

$$P_R = 8\text{ V} \times 2\text{ A} = 16\text{ W} \tag{2.13}$$

2.3 电阻的串并联组合

电路中的元件可以通过多种不同的方式连接,其中最简单的两种方式就是串联和并联。**串联**是将电路元件逐个顺次首尾相连接。因此,串联电路中通过所有元件的电流都是相同的。几个电路元件的两端分别连接于两个节点,此种连接方式称为**并联**。因此,并联电路中每个元件两端的电压相同。

下面,将从电阻、电容和电感三个方面去分析串并联组合。

1. 电阻

(1) 串联电阻的等效电阻如图 2.10 所示。

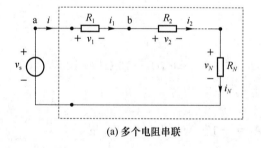 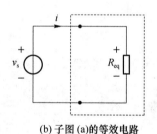

(a) 多个电阻串联　　　　　　　(b) 子图 (a) 的等效电路

图 2.10 串联电阻

如图 2.10(a) 所示,根据 KVL,可得

$$v_s = v_1 + v_2 + \cdots + v_N \tag{2.14}$$

根据欧姆定律,可以进一步得到

$$v_s = R_1 i_1 + R_2 i_2 + \cdots + R_N i_N \tag{2.15}$$

由于所有电阻都是串联的,所以电流为

$$i = i_1 = i_2 = \cdots = i_N \tag{2.16}$$

则

$$v_s = (R_1 + R_2 + \cdots + R_N)i \tag{2.17}$$

如果令 $v_s = R_{eq} i$,则

$$R_{eq} = \sum_{n=1}^{N} R_n \tag{2.18}$$

即所有电阻串联的总电阻值等于它们各自电阻的和。图 2.10(b)中的电路可以被视作图 2.10(a)中电路的等效电路。

电阻的串联在电路中十分常见。如图 2.11 所示的分压电路,就是电阻串联的一个经典应用实例。

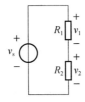

图 2.11 一个分压电路实例

根据 KVL、欧姆定律和串联电路特性,可知

$$\begin{cases} v_1 = \dfrac{R_1}{R_1 + R_2} v_s \\ v_2 = \dfrac{R_2}{R_1 + R_2} v_s \end{cases} \tag{2.19}$$

进一步扩展至 N 个串联电阻,则每个电阻的分压可记作

$$v_n = \frac{R_n}{\sum_{i=1}^{N} R_i} v_s \tag{2.20}$$

(2) 并联电阻的等效电阻

根据 KCL 和欧姆定律,图 2.12 中电路的电流为

$$i_s = i_1 + i_2 + \cdots + i_N = \frac{v}{R_1} + \frac{v}{R_2} + \cdots + \frac{v}{R_N} \tag{2.21}$$

如果令 $i_s = \dfrac{v}{R_{eq}}$,则

$$\frac{1}{R_{eq}} = \sum_{n=1}^{N} \frac{1}{R_n} \tag{2.22}$$

即并联电路中的总电阻的倒数等于各支路电阻的倒数和。图 2.12(b)中的电路可以被视作图 2.12(a)中的等效电路。

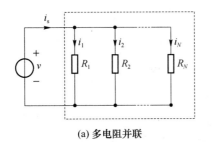

(a) 多电阻并联   (b) 子图 (a) 的等效电路

图 2.12 并联电阻

为了简化写法，下面引入一个新的概念——电导，用字母 G 表示，单位为西门子(S)。电导与电阻成反比，即 $G_n = \dfrac{1}{R_n}$。因此，电阻并联电路的等效电导可记作

$$G_{eq} = \sum_{n=1}^{N} G_n \tag{2.23}$$

电阻串联电路的等效电导可记作

$$\frac{1}{G_{eq}} = \sum_{k=1}^{N} \frac{1}{G_k} \tag{2.24}$$

图 2.13 所示的是一个双电阻并联电路，这两个电阻并联后的等效电阻为

$$R_{eq} = R_1 // R_2 = \frac{1}{\dfrac{1}{R_1} + \dfrac{1}{R_2}} = \frac{R_1 R_2}{R_1 + R_2} \tag{2.25}$$

注意，三个电阻并联后的等效电阻为

$$R_{eq} = R_1 // R_2 // R_3 = \frac{R_1 R_2 R_3}{R_1 R_2 + R_1 R_2 + R_2 R_3} \tag{2.26}$$

而不是 $\dfrac{R_1 R_2 R_3}{R_1 + R_2 + R_3}$。

电阻的并联在电路中十分常见。如图 2.13 所示的分流电路，就是电阻并联的一个经典应用实例。根据欧姆定律和并联电路特性，电压可以表示为

$$v = i_1 R_1 = i_2 R_2 = \frac{R_1 R_2}{R_1 + R_2} i_s \tag{2.27}$$

可得

$$\begin{cases} i_1 = \dfrac{R_2}{R_1 + R_2} i_s = \dfrac{G_1}{G_1 + G_2} i_s \\ i_2 = \dfrac{R_1}{R_1 + R_2} i_s = \dfrac{G_2}{G_1 + G_2} i_s \end{cases} \tag{2.28}$$

进一步扩展至 N 个并电阻，则每个电阻的分流可记作

$$i_n = \frac{G_n}{\sum_{i=1}^{N} G_i} i_s \tag{2.29}$$

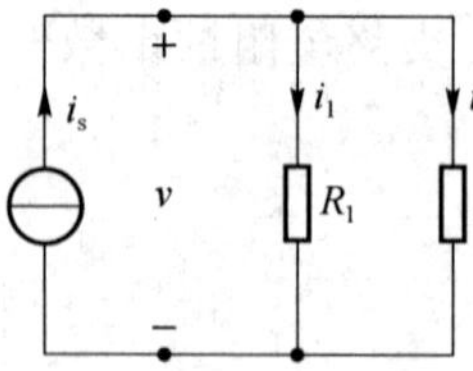

图 2.13 分流电路

问题 2.4 计算图 2.14 所示电路中的 v_o。

解：图 2.14 中的电路既不是串联，也不是并联，但它既包含了串联部分，也包含了并联部分。如图 2.14 所示，R_2 与 R_L 并联，其对应的等效电阻为

$$R_{eq} = \frac{R_2 R_L}{R_2 + R_L} \tag{2.30}$$

然后,类似于分压电路,可以得到

$$v_o = \frac{R_{eq}}{R_1 + R_{eq}} v_s \tag{2.31}$$

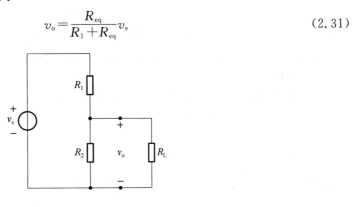

图 2.14　问题 2.4 中的电路

问题 2.5　计算图 2.15 所示电路中,a、b 两点之间的等效电阻 R_{ab}。

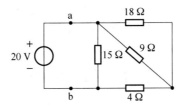

图 2.15　问题 2.5 中的电路

解:首先,为了使拓扑结构更加清晰,将图 2.15 中 a、b 两点之间的电路重绘为图 2.16 所示电路。

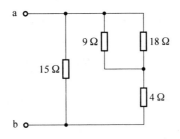

图 2.16　重绘图 2.15 所示电路

然后,根据电阻串并联组合的规律,得到等效电阻

$$R_{ab} = (9\,\Omega // 18\,\Omega + 4\,\Omega) // 15\,\Omega = (6\,\Omega + 4\,\Omega) // 15\,\Omega = 6\,\Omega \tag{2.32}$$

2. 电感

电感的串并联组合也遵循和电阻同样的规律,即串联的非耦合电感的总电感等于它们各自电感的和,记作

$$L_{eq} = \sum_{n=1}^{N} L_n \tag{2.33}$$

并联后的总电感的倒数等于它们各自电感的倒数之和,记作

$$\frac{1}{L_{eq}} = \sum_{n=1}^{N} \frac{1}{L_n} \tag{2.34}$$

式(2.33)和式(2.34)可以用 KCL、KVL 以及电感的 VCR 关系来证明。具体的证明方法可以在第 4 章学习完电感的 VCR 关系之后进行验证,此处只给出结论。

3. 电容

电容器的串并联组合与电感的串并联关系正好相反,即串联电容的总电容的倒数等于它们各自电容的倒数之和,记作

$$\frac{1}{C_{eq}} = \sum_{n=1}^{N} \frac{1}{C_n} \tag{2.35}$$

并联后的总电容等于它们各自的电容之和,记作

$$C_{eq} = \sum_{n=1}^{N} C_n \tag{2.36}$$

式(2.35)和式(2.36)可以用 KCL、KVL 以及电容的 VCR 关系来证明。具体的证明方法可以在第 4 章学习完电容的 VCR 关系之后进行验证,此处只给出结论。

综上所述,电路元件的等效变换是电路分析中的一种简化电路分析过程的有效方法。

2.4 2b 方法

本节在 KCL、KVL 和欧姆定律的基础上,介绍一种通用的电路分析方法。假设一个电路里有 n 个节点和 b 条支路,则可以形成 $m=b-(n-1)$ 个网孔。根据 KCL,可以对任意 $(n-1)$ 个节点列出 $(n-1)$ 个独立方程;根据 KVL,可以对 m 个网孔列出 $b-(n-1)$ 个独立方程;根据欧姆定律,可以对 b 个支路列出 b 个 VCR 方程。因此,就可以得到 $(n-1)$ 个独立的 KCL 方程,$b-(n-1)$ 个独立的 KVL 方程,b 个元件的 VCR 方程。把这些方程整合起来,可以得到 $2b$ 个独立的方程,就可以解出 $2b$ 个未知变量。这些变量就是电路中 b 个元件的电压值和电流值,这种方法就称为 $2b$ 法。需要注意的是,如果电路中含有 s 个独立的电源,则电路中的未知变量的数量变为 $s+2(b-s)=2b-s$。

问题 2.6 采用 $2b$ 法计算图 2.17 所示电路中的所有电流值和电压值。

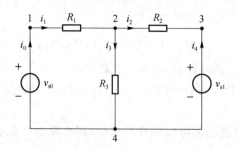

图 2.17 问题 2.6 中的电路

解:图 2.17 中包含 $n=4$ 个节点和 $b=5$ 条支路。因此,可对图中 $n-1=3$ 个节点(如节点 1、2、3)列出 3 个独立的 KCL 方程。

$$\begin{cases} i_0 - i_1 = 0 \\ i_1 - i_2 - i_3 = 0 \\ i_2 + i_4 = 0 \end{cases} \tag{2.37}$$

注:节点4的KCL方程$(-i_0 + i_3 - i_4 = 0)$可以由节点1、2、3的方程推出。因此,它不是一个独立的方程。

对图中的$b-(n-1)=2$个网孔(图2.18中的网孔Ⅰ和网孔Ⅱ)列出2个独立的KVL方程,

$$\begin{cases} v_1 + v_3 - v_{s0} = 0 \\ -v_3 + v_2 + v_{s1} = 0 \end{cases} \tag{2.38}$$

此外,对于$b=5$条支路,列出VCR方程

$$\begin{cases} v_1 = R_1 i_1 \\ v_2 = R_2 i_2 \\ v_3 = R_3 i_3 \\ v_{s0} = \text{given} \\ v_{s1} = \text{given} \end{cases} \tag{2.39}$$

将上述列出的10个独立方程进行整合即可求出电路中所有的电流值和电压值。

$2b$法的求解步骤总结如下:

第一步:对任意$(n-1)$个节点列出$(n-1)$个独立方程;根据KVL,可以对m个网孔列出$b-(n-1)$个独立方程;根据欧姆定律,对b个支路列出b个VCR方程。

第二步:求解上述方程,可得到电路中多个元件的电压和电流值。

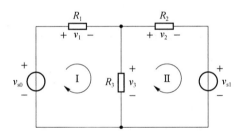

图2.18 图2.17中的电路标出网孔

2.5 节点电压法

2.4节给出了电路分析中的一种通用方法——$2b$法。采用$2b$法可以求出电路中所有电压和电流变量。然而,在某些情况下,并不需要求解所有变量,只需求解部分变量。因此,在本节和2.6节,将介绍两种快速求解特定变量的电路分析方法,分别为节点电压法和网孔电流法。本节首先介绍节点电压法的求解过程。

节点电压法是以节点电压为电路变量列写方程进行求解的一种分析方法。节点电压是其相对于参考节点电压的电压升高值。它的基础是KCL和元件的VCR关系,其求解基本思路为:

(1) 选择参考节点并定义节点电压；
(2) 为除参考节点外的节点构建 KCL 方程；
(3) 求解方程得到各个节点电压；
(4) 通过节点电压，求解其他所需的未知变量。

问题 2.7 用节点电压法求图 2.19 中的 v_1、v_2 和 i_1。

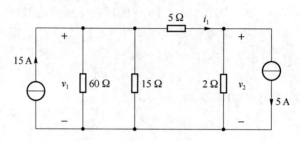

图 2.19　问题 2.7 中的电路

解：选择参考节点定义节点电压，如图 2.20 所示。

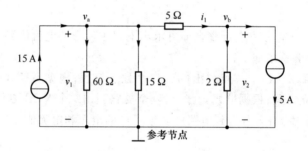

图 2.20　图 2.19 所示电路加入参考节点

为除参考节点外的节点构建 KCL 方程

$$\begin{cases} 15\ \text{A} - \dfrac{v_a}{60\ \Omega} - \dfrac{v_a}{15\ \Omega} - \dfrac{v_a - v_b}{5\ \Omega} = 0 \\ \dfrac{v_a - v_b}{5\ \Omega} - \dfrac{v_b}{2\ \Omega} - 5\ \text{A} = 0 \end{cases} \tag{2.40}$$

求解方程得到各节点电压

$$\begin{cases} v_a = 60\ \text{V} \\ v_b = 10\ \text{V} \end{cases} \tag{2.41}$$

通过节点电压确定所求未知变量 $v_1 = v_a = 60\ \text{V}, v_2 = v_b = 10\ \text{V}, i_1 = \dfrac{v_a - v_b}{5} = \dfrac{60\ \text{V} - 10\ \text{V}}{5\ \Omega} = 10\ \text{A}$。

问题 2.8 用节点电压法求图 2.21 中的节点电压 v_1、v_2、v_3 和 v_4。

解：选择参考节点，并对其他各个节点构建 KCL 方程

$$\begin{cases} 1 - 0.1v_1 - 0.1(v_1 - v_4) - (v_1 - v_2) = 0 \\ (v_1 - v_2) - v_2 - 0.5(v_2 - v_3) - 0.5 = 0 \\ 0.5(v_2 - v_3) + 0.5 - 0.5v_3 - 0.25(v_3 - v_4) = 0 \\ 0.1(v_1 - v_4) + 0.25(v_3 - v_4) - 0.25v_4 = 0 \end{cases} \tag{2.42}$$

求解方程得到各节点电压

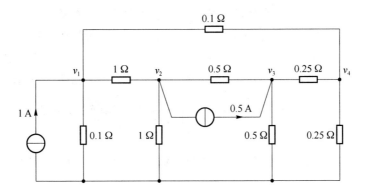

图 2.21 问题 2.8 中的电路

$$\begin{cases} v_1 = 1.23 \text{ V} \\ v_2 = 0.42 \text{ V} \\ v_3 = 0.67 \text{ V} \\ v_4 = 0.48 \text{ V} \end{cases} \tag{2.43}$$

注意：如果支路中有电压源（无论是独立电压源还是受控电压源），流过它的电流无法用节点电压来表示。因此，在电压源支路两端的两个节点处均无法构建 KCL 方程。在这种情况下，需要根据以下两种情况进行处理：

(1) 如果电压源支路中的一个节点为参考节点，则另一个节点的电压便是已知的（对于独立电压源）或者说是可以被表示出来的（对于受控电压源）。因此无须为该节点建立 KCL 方程。

(2) 如果电压源两端都不是参考节点，则需要列出两个独立方程才能符合求解要求。其中一个是超节点的 KCL 方程，另一个是电压源支路的电压差方程。

下面介绍什么是超节点。比如将电压源及其所在支路的两端的两个节点组合成一个节点，那么该节点就是超节点。每个超节点包含两个节点，一个是非参考节点，另一个既可以是非参考节点也可以是参考节点。根据超节点的定义，图 2.21 中的受控电压源、节点 1 和节点 4 就构成一个超节点。根据 KCL 的定义，KCL 对于超节点也成立。

问题 2.9 用节点电压法求 v_3 和 v_4 的值。

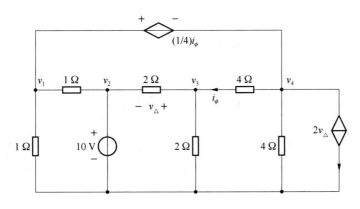

图 2.22 问题 2.9 中的电路

解:选取参考节点,列出其他节点的 KCL 方程,并为超节点构建 KCL 和电压差分方程

$$\begin{cases} 2(v_2-v_3)-2v_3+4(v_4-v_3)=0 &\text{------节点 3 的 KCL 方程} \\ v_2=10 &\text{------节点 2 的 KCL 方程} \\ -v_1+(v_2-v_1)-4(v_4-v_3)-4v_4-2(v_3-v_2)=0 &\text{----超节点的 KCL 方程} \\ v_1-v_4=\dfrac{1}{4}\times 4(v_4-v_3) &\text{------电压源所在支路的电压差方程} \end{cases} \quad (2.44)$$

求解方程组得 $v_3=4.5\text{ V}, v_4=4\text{ V}$。

问题 2.10 如图 2.23 所示,R_2 的功率为 2 W,求解 R_1、R_2 和 R_3 的阻值。

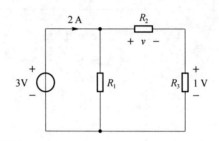

图 2.23 问题 2.10 中的电路

解:用 KVL 计算 R_2 的电压值,

$$v=3\text{ V}-1\text{ V}=2\text{ V} \quad (2.45)$$

然后反推出 R_2 的阻值,

$$R_2=\frac{v^2}{P}=\frac{(2\text{ V})^2}{2\text{ V}\cdot\text{A}}=2\text{ }\Omega \quad (2.46)$$

利用欧姆定律,通过电阻 R_2 和 R_3 的电流值为

$$I_2=\frac{v}{R_2}=\frac{2\text{ V}}{2\text{ }\Omega}=1\text{ A} \quad (2.47)$$

$$R_3=\frac{1}{I_2}=\frac{1\text{ V}}{1\text{ A}}=1\text{ }\Omega \quad (2.48)$$

根据 KCL,通过 R_1 的电流为

$$I_1=2\text{ A}-I_2=2\text{ A}-1\text{ A}=1\text{ A} \quad (2.49)$$

进一步推出

$$R_1=\frac{3\text{ V}}{I_1}=\frac{3\text{ V}}{1\text{ A}}=3\text{ }\Omega \quad (2.50)$$

2.6 网孔电流法

本节介绍第二种快速求解特定变量的电路分析方法——网孔电流法。网孔电流法是以网孔电流(而不是支路电流)为电路变量列方程进行求解的电路分析方法。本节之前所涉及的电流均为支路电流。支路电流是通过电路的实际电流,而网孔电流是一种假想的沿着网孔边界流动的电流。电路中,假设每个网孔只有一个网孔电流。图 2.24 表示了网孔电流与支路电流的关系。

图 2.24 (a) 中的电流 i_a、i_b 和 i_c 是支路电流,而图 2.24(b) 中的电流 i_{m1} 和 i_{m2} 是网孔电流。它们之间的关系为 $i_a = i_{m1}$,$i_b = i_{m2}$ 和 $i_c = i_{m1} - i_{m2}$。

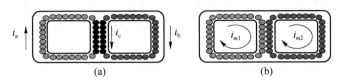

图 2.24 网孔电流与支路电流的关系

网孔电流法的基础是 KVL 和元件的 VCR 关系,其求解基本思路为:
(1) 规定每个网孔中的网孔电流及其参考方向;
(2) 为每个网孔构建 KVL 方程;
(3) 求解网孔电流方程;
(4) 通过网孔电流,求解其他所需的未知变量。

问题 2.11 在图 2.25 中,$R_1 = 5\ \Omega$,$R_2 = 10\ \Omega$,$R_3 = 20\ \Omega$,$v_{s0} = 20\ \text{V}$,$v_{s1} = 10\ \text{V}$,计算每个支路的电流。

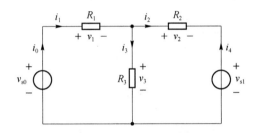

图 2.25 问题 2.11 中的电路

解:如图 2.26 所示,设定网孔电流 i_a 和 i_b。

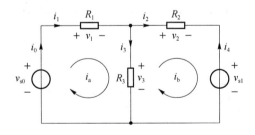

图 2.26 图 2.25 中电路设定网孔电流

列出每个网孔的 KVL 方程

$$\begin{cases} R_1 i_a + R_3 (i_a - i_b) - v_{s0} = 0 \\ R_2 i_b + v_{s1} + R_3 (i_b - i_a) = 0 \end{cases} \tag{2.51}$$

求解方程组,得到网孔电流 i_a 和 i_b

$$i_a = \frac{8}{7}\ \text{A},\ i_b = \frac{3}{7}\ \text{A} \tag{2.52}$$

计算支路电流

$$i_0 = i_1 = i_a = \frac{8}{7}\text{A}, i_2 = i_b = \frac{3}{7}\text{A}, i_3 = i_a - i_b = \frac{5}{7}\text{A} \text{ 和 } i_4 = -i_b = -\frac{3}{7}\text{A}。$$

注意：如果支路中含有电流源(无论是独立电流源还是受控电流源)，它两端的电压不能用网孔电流表示。因此，有电流源支路的网孔无法构建 KVL 方程。在这种情况下，需要根据以下两种情况进行处理：

(1) 如果含有电流源的支路位于电路边缘，则其对应的网孔电流已知且等于该支路电流(对于独立电流源)，或说可以被表示出来(对于受控电流源)。因此无须为该网孔建立 KVL 方程。

(2) 如果含有电流源的支路不在电路边缘，则有两个包括了该电流源支路的网孔，因此需要建立两个方独立程。其中，一个是超网孔的 KVL 方程，另一个是电流源支路的电流差方程。

下面介绍什么是超网孔。超网孔是一个包括了两个网孔的回路。当一个电流源被两个网孔包括时，就会生成一个超网孔。此时，可得到了一个包含两个网孔电流的 KVL 方程。而另一个方程表示电流源等于一个网孔电流减去另一个网孔电流。

问题 2.12 用网孔电流法求图 2.27 中电路的总功耗。

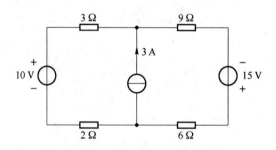

图 2.27 问题 2.12 中的电路

解：选择图 2.26 中的网孔电流 i_a 和 i_b，此外虚线表示超网孔。

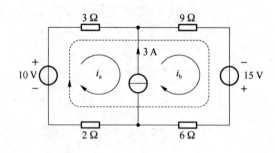

图 2.28 图 2.27 中电路加上网孔电流

建立电流差方程和超网孔 KVL 方程

$$\begin{cases} i_b - i_a = 3 \text{ A} \\ i_a(3\,\Omega + 2\,\Omega) + i_b(9\,\Omega + 6\,\Omega) - 15\text{ V} - 10\text{ V} = 0\text{ V} \end{cases} \quad (2.53)$$

得到 $i_a = -1$ A 和 $i_b = 2$ A。

因此，所有电阻消耗的功率是 $i_a^2(3\,\Omega + 2\,\Omega) + i_b^2(9\,\Omega + 6\,\Omega) = 65$ W，三个电源产生的功

率是 $-10i_a-15i_b+3(15-6i_b-9i_b)=-65$ W。

在分析节点电压法和网孔电流法的基础上,对这两种方法进行比较:

(1) 网孔电流分析只适用于平面电路;

(2) 节点电压法可直接得到电压值,而网孔电流法得到的是网孔电流值而非支路电流值;

(3) 网孔电流法求解 $m=b-(n-1)$ 个方程,而节点电压法求解 $(n-1)$ 个方程。其中,$m=b-(n-1)<b<2b$ 和 $n-1<2b$ 恒成立。所以,利用节点电压法和网孔电流均可以减少待解方程的数量,简化电路分析过程。

2.7 习　　题

习题 2.1　图 2.29 中有多少分支、节点和网孔?

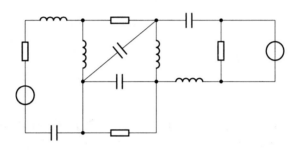

图 2.29　习题 2.1 中的电路

习题 2.2　计算图 2.30(a)和(b)中的等效电阻。

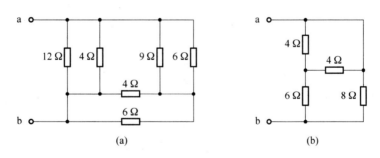

图 2.30　习题 2.2 中的电路

习题 2.3　计算图 2.31 中的等效电阻。

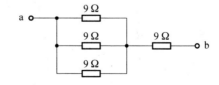

图 2.31　习题 2.3 中的电路

习题 2.4　计算图 2.32 中的 v_1、v_2 和 i_1。

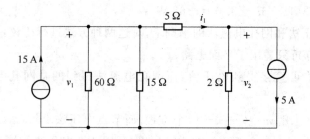

图 2.32 习题 2.4 中的电路

习题 2.5 电路如图 2.33 所示。(a)计算 i_a，i_b 和 v_o 的值；(b) 计算每个电阻消耗的功率和电压源提供的功率。

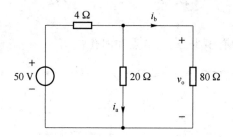

图 2.33 习题 2.5 中的电路

习题 2.6 图 2.34 电路中的电流 i_o 为 4 A。(a)计算 i_1；(b)计算每个电阻消耗的功率；(c)验证电路中消耗的总功率等于 180 V 电源产生的功率。

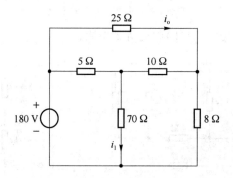

图 2.34 习题 2.6 中的电路

习题 2.7 计算图 2.35 中的 i_2，i_1 和 i_o。

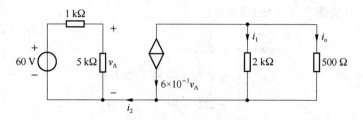

图 2.35 习题 2.7 中的电路

习题 2.8　计算图 2.36 中的 v_1 和 v_2。

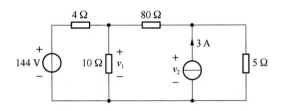

图 2.36　习题 2.8 中的电路

习题 2.9　计算图 2.37 中的 v_o。

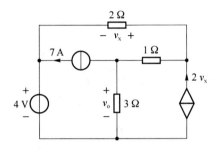

图 2.37　习题 2.9 中的电路

习题 2.10　求出图 2.38 所示电路的总功率,并验证求解后的答案是否符合产生的总功率等于消耗的总功率这一原则。

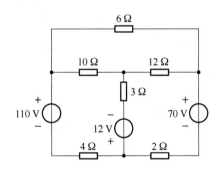

图 2.38　习题 2.10 中的电路

习题 2.11　对图 2.39 中电路法求解 i_Δ。计算出独立电流源的输出功率,并计算受控电压源产生的功率。

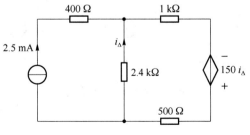

图 2.39　习题 2.11 中的电路

习题 2.12 求出图 2.40 所示电路的总功率。

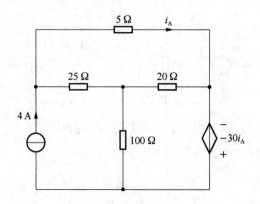

图 2.40 习题 2.12 中的电路

习题 2.13 在图 2.41 中,电阻 R 的电压值为 $v=8\text{ V}$,计算 R 的阻值。

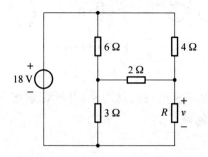

图 2.41 习题 2.13 中的电路

习题 2.14 在图 2.42 所示电路中,使用网孔电流法计算支路电流 $i_1 \sim i_6$。

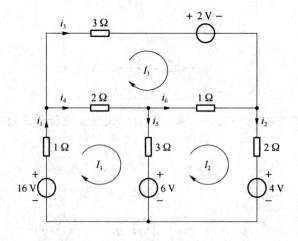

图 2.42 习题 2.14 中的电路

习题 2.15 对图 2.43 所示电路使用网孔电流法,求出电路中的电流 I。

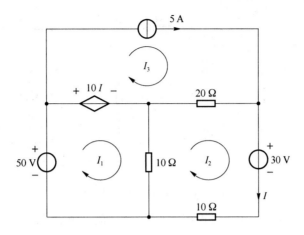

图 2.43 习题 2.15 中的电路

习题 2.16 对图 2.44 所示电路使用网孔电流法,求出电路中的电压 v。

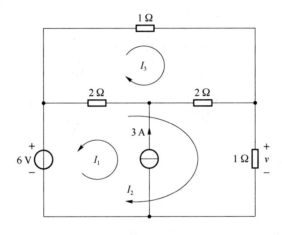

图 2.44 习题 2.16 中的电路

习题 2.17 在图 2.45 电路中,$v_{ab}=5$ V,求出电路中的电压 v_s。

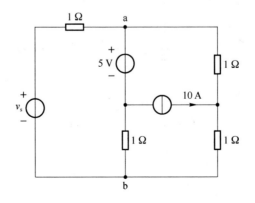

图 2.45 习题 2.17 中的电路

习题 2.18　使用节点电压法求出图 2.46 电路中的电压 v_{ab}。

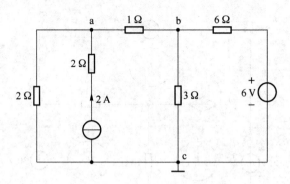

图 2.46　习题 2.18 中的电路

习题 2.19　使用节点电压法求出图 2.47 电路中的电流 i_1 和 i_2。

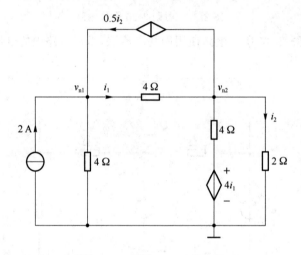

图 2.47　习题 2.19 中的电流

习题 2.20　求题图 2.48 所示电路中的电流 i_1、i_2 和 i_3。

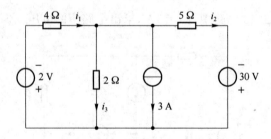

图 2.48　习题 2.20 中的电流

第 3 章

电路基本理论

第 2 章介绍了电路理论中的基本分析方法,这些方法可直接分析简单的电路。但随着电路结构变得更加复杂,希望可以找到简化的方法,将复杂电路进行拆分、合并或是等效。因此,本章将介绍几个电路基本定理,用于复杂电路的简化。

第 3 章的学习目标:电路基本定理

(1) 理解并掌握叠加定理;

(2) 理解并掌握电压源电流源的等效变换;

(3) 理解并掌握戴维南定理和诺顿定理;

(4) 理解并掌握最大功率传输定理。

3.1 叠 加 定 理

在由线性电阻、线性受控源和独立源组成的电路中,每个元件的电压或电流可以看成每个独立源单独作用于电路时,在该元件产生的电压或电流的代数和,这就是电路的叠加定理。

"单独作用"意味着当某一个独立源单独作用时,其他独立源应为 0,即独立电压源短路,独立电流源断路。

注意:

(1) 根据受控源的特点,其是不能单独作用的,所以电路中不能单独保留受控源。但在独立源单独作用时,受其控制的受控源需要保留在电路中。

(2) 叠加定理不适用于功率的计算。

问题 3.1 利用叠加定理计算电流 i_1 和 i_2。

解: 图 3.1 电路中有两个独立的电源。根据叠加定理,电路可分为两个子电路,如图 3.2(a)和(b)所示。下面分别分析这两个子电路。

对于图 3.2(a)中的子电路

$$i_1^{(1)} = i_2^{(1)} = \frac{54 \text{ V}}{9 \text{ }\Omega + 8 \text{ }\Omega} = \frac{54 \text{ V}}{27 \text{ }\Omega} = 2 \text{ A} \tag{3.1}$$

对于图 3.2(b)中的子电路

$$i_1^{(2)} = -\frac{18\ \Omega}{9\ \Omega + 18\ \Omega} \times 6\ A = -4\ A \tag{3.2}$$

$$i_2^{(2)} = \frac{9\ \Omega}{9\ \Omega + 18\ \Omega} \times 6\ A = 2\ A \tag{3.3}$$

然后,根据叠加定理,原图的电流是两个子电路中电流的代数和

$$i_1 = i_1^{(1)} + i_1^{(2)} = 2\ A - 4\ A = -2\ A \tag{3.4}$$

$$i_2 = i_2^{(1)} + i_2^{(2)} = 2\ A + 2\ A = 4\ A \tag{3.5}$$

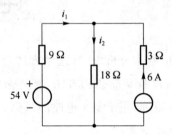

图 3.1 问题 3.1 中的电路

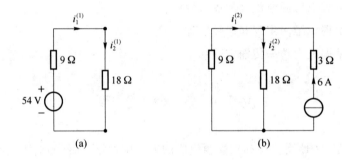

图 3.2 图 3.1 的两个子电路

问题 3.2 利用叠加定理计算图 3.3 中的电压 v。

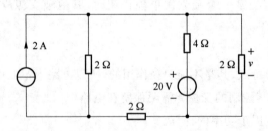

图 3.3 问题 3.2 中的电路

解:图 3.3 电路中有两个独立的电源。根据叠加定理,电路可以划分为两个子电路如图 3.4(a)和(b)所示。下面分别分析这两个子电路。

分别求解得到两个子电路的电压 $v^{(1)} = 1$ V 和 $v^{(2)} = 5$ V。因此,原图的电压为

$$v = v^{(1)} + v^{(2)} = 1\ V + 5\ V = 6\ V \tag{3.6}$$

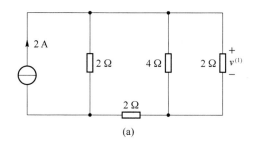
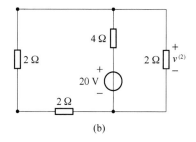

图 3.4　图 3.3 的两个子电路

3.2　电压源与电流源的等效变换

在电路中,电源像一个黑盒子,电源两端电压与通过电源的电流的关系描述了电源的所有性质。如果两个电源有相同的电压电流函数关系,那么就可以认为这两个电源相同。

如图 3.5(a)所示,电压源可以抽象为一个提供固定电压的元件与一个电阻串联,此时电压源对外电路提供的电压为

$$v = v_s - iR_s \tag{3.7}$$

其中,v_s 是开路电压,R_s 是串联电阻。

如图 3.5(b)所示,电流源可以抽象为一个提供固定电流的元件与一个电阻并联,此时电流源对外电路提供的电流为

$$i = i_s - \frac{v}{R_p} \tag{3.8.a}$$

或

$$v = i_s R_p - i R_p \tag{3.8.b}$$

其中,i_s 是短路电流,R_p 是并联电阻。

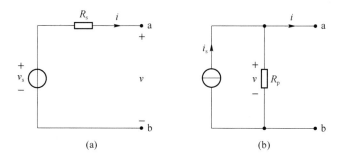

图 3.5　电压源电流源的等效变换

观察式(3.7)和式(3.8)可知,如果满足

$$\begin{cases} R_s = R_p \\ v_s = i_s R_p \end{cases} \tag{3.9}$$

则图 3.5(a)和图 3.5(b)中的电压源和电流源可为外电路提供相同电压和电流,即该电压源和电流源等价。

电压源电流源的等效变换是指一个与某个电阻串联的电压源可以等效为一个与同一电阻并联的电流源,反之亦然。电压源电流源等效变换需要满足的条件如式(3.9)所示。

注意:

(1) 电压源和电流源的等效关系只对外电路而言,对电源内部是不等效的。

(2) 等效变换时,两电源的参考方向要对应,如图 3.6(a)和图 3.6(b)所示。

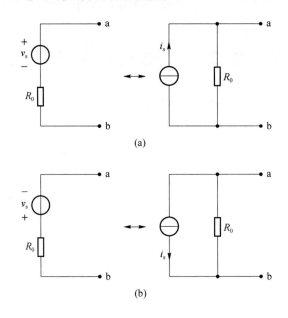

图 3.6 电压源电流源的等效变换的参考方向

此外,如果有一个电阻与电压源并联,或者有一个电阻与电流源串联,则电阻对端子 a 和 b 的等效电路没有影响,如图 3.7(a)和(b)所示。

电压源电流源的等效变换并不局限于电阻电路,在频域中通过将电路元件表示为阻抗和电源,也可以将电压源电流源的等效变换应用在包含电容和电感的电路中。

问题 3.3 在图 3.8 所示的电路中,使用电压源电流源的等效变换计算电流 i。

解: 根据电压源电流源的等效变换,可将图 3.8 中的电路重绘为图 3.9 所示的电路。因此,电流为

$$i = \frac{12\text{ V} - 8\text{ V}}{4\text{ }\Omega + 2\text{ }\Omega + 2\text{ }\Omega} = \frac{4\text{ V}}{8\text{ }\Omega} = 0.5 \text{ A} \tag{3.10}$$

问题 3.4 使用电压源电流源的等效变换,求图 3.10 中 5 kΩ 电阻中的电流。

解: 通过电压源电流源的等效变换,可得到从图 3.11(a)到(g)的变换。然后便得到

$$i_0 = \frac{12\text{ V}}{35\text{ k}\Omega + 5\text{ k}\Omega} = 3 \text{ mA}$$

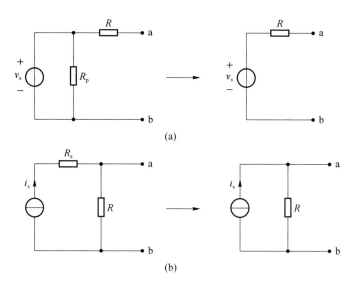

(a)

(b)

图 3.7 电压源电流源等效变换中的特殊情况

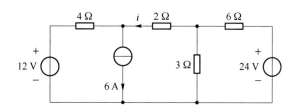

图 3.8 问题 3.3 中的电路

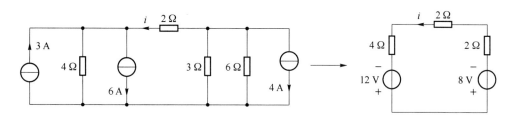

图 3.9 图 3.8 中电路的等效电路

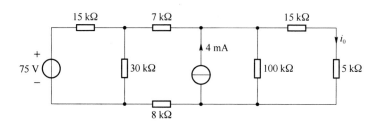

图 3.10 问题 3.4 中的电路

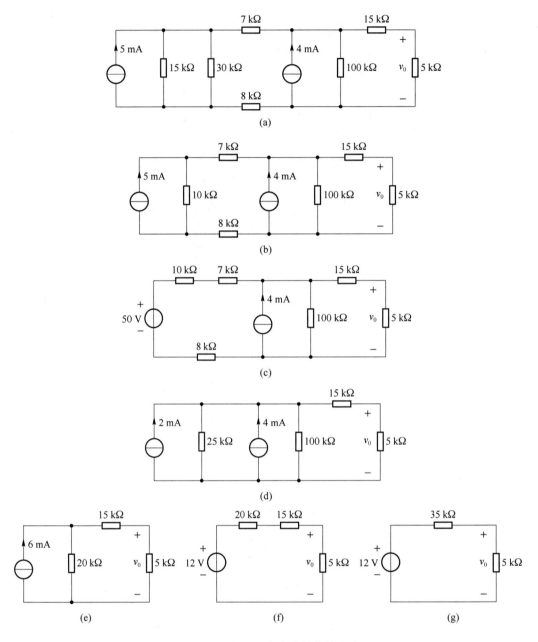

图 3.11　图 3.10 中电路的等效电路

3.3　戴维南定理和诺顿定理

在电路分析中,很多时候并不需要分析电路中的每个参数,可能只要获知特定网络端口提供的电压和电流。例如,手机充电时,我们关心的是电源给充电接头提供的电压和电流,而电源电路内部的电压或电流我们并不关心。所以,可将电源内部电路进行简化。

在第 2 章,已经介绍了串联电路和并联电路的等效方法,但是实际电路都是由电阻、电源通过串并联混合连接在一起的,除按步进行串并联转换外,是否有其他更加简单的等效方法?

戴维南定理和诺顿定理是用来分析电路中端口网络的电路简化技术。

首先介绍戴维南定理(又称等效电压源定律),其内容是:一个含有独立电源和电阻的线性网络的两端,其对外部而言可以用一个独立电压源串联一个电阻来等效。其中,电压源的电压值 v_{Th} 等于 ab 两端开路时的电压 v_{oc},串联电阻的阻值 R_{Th} 等于网络内全部独立电源置零时网络 N 的 ab 两端的等效电阻。如图 3.12 所示,是一个网络 ab 两端的戴维南等效电路。无论网络 N 的具体电路连接结构和参数如何设计,只要它们可以等效为相同的戴维南等效电路,它们在 ab 端为外电路提供电压和电流的能力就相同;当它们连接相同的负载时,在负载上可得到相同的电压和电流。

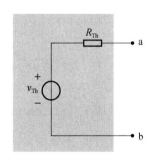

图 3.12 电路网络 N

问题 3.5 画出图 3.13 中 ab 端的戴维南等效电路。

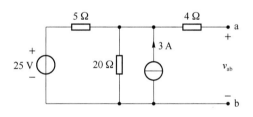

图 3.13 问题 3.5 中的电路

解:求 ab 两端间的开路电压。

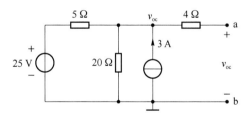

图 3.14 ab 端开路后,问题 3.5 中的电路

通过节点电压分析方法,得到

$$\frac{v_{oc}-25\text{ V}}{5\text{ }\Omega}+\frac{v_{oc}}{20\text{ }\Omega}-3\text{ A}=0\text{ A} \tag{3.11}$$

求得开路电压 $v_{oc}=32$ V。因此,戴维南等效电路中的电压源电压值 $v_{Th}=v_{oc}=32$ V。

然后,去掉电路中所有电源(如图 3.15 所示),求出 ab 端间的戴维南等效电阻。

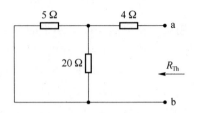

图 3.15 问题 3.5 中电路去掉电路中所有电源,ab 两端的等效电阻

由图 3.15 可知,ab 两端的戴维南等效电阻 $R_{Th}=8\ \Omega$。因此,ab 端的戴维南等效电路如图 3.16 所示。

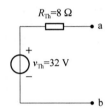

图 3.16 问题 3.5 中电路的戴维南等效电路

从图 3.12 所示的戴维南等效电路,结合 3.2 节中介绍的电压源电流源的等效变换,可以将一个电压源与一个电阻的串联等效为一个电流源与一个电阻的并联,即图 3.12 可以等效为图 3.17。其中,电流源电流值和并联电阻值也可以根据式(3.9)求出。

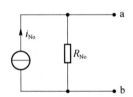

图 3.17 诺顿定理示意图

此时,可以得到诺顿定理。

诺顿定理又称等效电流源定律,其内容是:一个含有独立电源和电阻的线性网络的两端,其对外部而言可以用一个独立电流源并联一个电阻来等效。其中,电流源的电流值 i_{No} 等于 ab 两端短路时的电流 i_{sc},并联电阻的阻值 R_{No} 等于网络内全部独立电源置零时网络 N 的 ab 两端的等效电阻。且由电压源电流源等效变化可知

$$\begin{cases} R_{No}=R_{Th} \\ i_{sc}=\dfrac{v_{Th}}{R_{Th}} \end{cases} \quad (3.12)$$

与戴维南定理类似,可得无论网络 N 的具体电路连接结构和参数如何设计,只要它们可以等效为相同的诺顿等效电路,它们在 ab 端为外电路提供电压和电流的能力就相同;当它们连接相同的负载时,在负载上可得到相同的电压和电流。

利用戴维南定理或诺顿定理,可将复杂的线性电路(不含负载部分)进行简化,降低电路

分析的难度。同时,这两个定理在交流电路中的应用与在直流电路中的应用完全相同,只是电阻被推广为阻抗。

3.4 最大功率传输定理

在实际的电路应用中,我们最关心的是电路是否可以为负载提供足够的功率,以支持负载正常工作。也就是说我们希望电路中电源所消耗的能量尽可能多地提供给负载。因此,本节针对电路中的功率传输问题,讨论了负载大小与功率传输之间的关系。

基于3.3节的分析,任何含有独立电源的线性单口网络可以表示为图3.18中的电路模型,其中R_L是ab端口网络的负载。此时,传输给负载R_L的功率为

$$P=\left(\frac{v_{Th}}{R_{Th}+R_L}\right)^2 R_L \qquad (3.13)$$

为求得负载R_L为多大时,令$\frac{dP}{dR_L}=0$,可获得最大的功率。可得当$R_L=R_{Th}$时,负载R_L获得最大的功率

$$P_{max}=\left(\frac{v_{Th}}{R_{Th}+R_L}\right)^2 R_L=\left(\frac{v_{Th}}{R_{Th}+R_L}\right)^2 R_{Th}=\frac{v_{Th}^2}{4R_L} \qquad (3.14)$$

上述结论被记作**最大功率传输定理**。该定理揭示了当电源电路确定后,获得最大功率所需的负载值,即当负载阻值等于供电电路内阻时,负载可获得最大的功率。

如果负载的电阻大于电源的电阻,电源传输到负载上的功率占比会变高,但由于电路中总电阻增加,负载的功率会降低;如果负载的电阻小于电源的电阻,虽然总电阻减小,但大部分的功率会在电源内部消耗,此时负载获得的功率减小。

该定理告诉我们当电源电阻给定后,应该如何选择负载电阻才能实现最大功率传输。然而,该定理并没有说明如何为给定的负载电阻选择电源电阻!人们往往会对该定理有一种误解,就是选择与负载值相同的电源阻值可以实现负载功率的最大化。事实上,这个结论显而易见是错误的。因为无论负载电阻是多少,使功率传输最大化的电源电阻总是零。

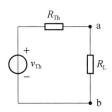

图 3.18 最大功率传输电路

最大传输定理同样也可以推广到交流电路中,即当负载阻抗等于电源阻抗的复共轭时,可以实现最大功率传输,该结论的证明会在第5章给出。

3.5 习　　题

习题 3.1　使用叠加定理计算图 3.19 所示电路中的电压 v。

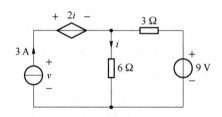

图 3.19　习题 3.1 中的电路

习题 3.2　使用叠加定理计算图 3.20 中的电流 i。

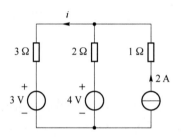

图 3.20　习题 3.2 中的电路

习题 3.3　图 3.21 所示电路中,已知 $v_s = 20$ V,求：
　　(a) 电流 i;
　　(b) 若要使 $i = 0.25$ A,v_s 应为多少。

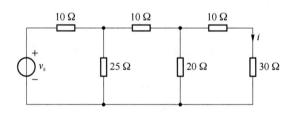

图 3.21　习题 3.3 中的电路

习题 3.4　利用叠加定理求图 3.22 所示电路中的电流 i。

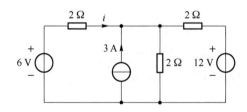

图 3.22　习题 3.4 中的电路

习题 3.5　使用电压源电流源等效变换求图 3.23 所示电路中的 i_1 和 i_2，然后计算阻值为 18 Ω 的电阻的功率和电压源的功率。

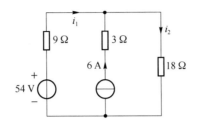

图 3.23　习题 3.5 中的电路

习题 3.6　使用电压源电流源等效变换计算图 3.24 所示电路中的电流 i。

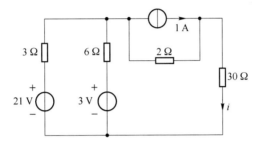

图 3.24　习题 3.6 中的电路

习题 3.7　电压源电流源等效变换计算图 3.25 所示电路中的电压 v_{ab}。

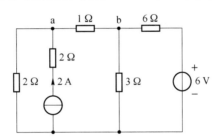

图 3.25　习题 3.7 中的电路

习题 3.8　根据电压源电流源等效变换计算图 3.26 所示电路中的电流 i_0。

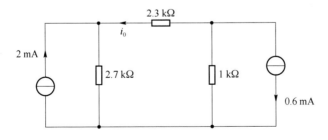

图 3.26　习题 3.8 中的电路

习题 3.9　根据电压源电流源等效变换计算图 3.27 所示电路中的电流 i_0。

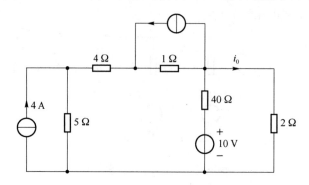

图 3.27　习题 3.9 中的电路

习题 3.10　根据叠加定理求出图 3.28 所示电路中的电压 v_0。

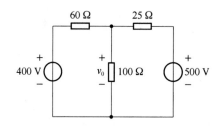

图 3.28　习题 3.10 中的电路

习题 3.11　根据叠加定理求出图 3.29 所示电路中的电压 v_0。

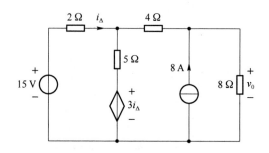

图 3.29　习题 3.11 中的电路

习题 3.12　画出图 3.30 电路中 ab 端对应的戴维南等效电路。

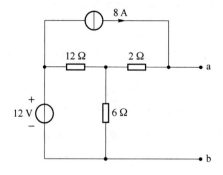

图 3.30　习题 3.12 中的电路

习题 3.13 画出图 3.31 电路中 ab 端对应的戴维南等效电路。

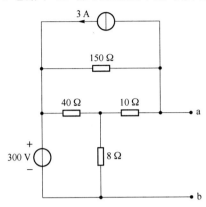

图 3.31 习题 3.13 中的电路

习题 3.14 画出图 3.32 电路中 ab 端对应的诺顿等效电路。

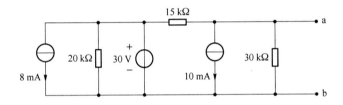

图 3.32 习题 3.14 中的电路

习题 3.15 画出图 3.33 电路中 ab 端对应的戴维南等效电路。

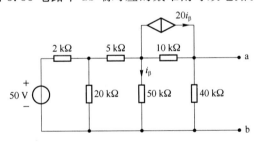

图 3.33 习题 3.15 中的电路

习题 3.16 画出图 3.34 电路中 ab 端对应的戴维南等效电路。

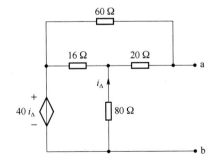

图 3.34 习题 3.16 中的电路

习题 3.17　图 3.35 中的电路通过调整可变电阻 R_0 实现了最大功率传输，计算 R_0 为何值时可以获得最大功率，求出 R_0 能获得的最大功率的值。

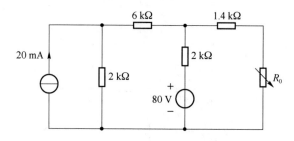

图 3.35　习题 3.17 中的电路

习题 3.18　求图 3.36 所示电路 ab 端的戴维南等效电路。

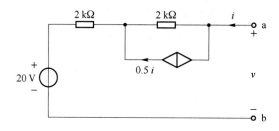

图 3.36　习题 3.18 中的电路

习题 3.19　求图 3.37 所示电路 ab 端的诺顿等效电路。

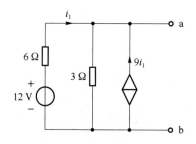

图 3.37　习题 3.19 中的电路

习题 3.20　图 3.38 所示电路中，N 为线性含源电阻网络。已知当开关 S1、S2 均打开时，电压表的读数为 6 V；当开关 S1 闭合 S2 打开时，电压表的读数为 4 V。试求当开关 S1 打开 S2 闭合时电压表的读数。

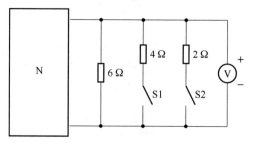

图 3.38　习题 3.20 中的电路

习题 3.21　试求图 3.39 所示电路中 R_L 为何值时可获得最大功率,最大功率是多少?

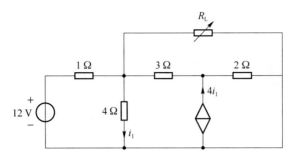

图 3.39　习题 3.21 中的电路

第 4 章

RC电路和RL电路

在前几章中,学习了电路分析中的基本方法和基本定理,并用这些知识分析了一些简单的电路。这些电路尽管连接拓扑不一样,但是均由电阻和电源组成。根据第 1 章所学的知识,我们知道电阻是静态元件。而在本章中,将对一类含有动态元件(电感和电容)的电路进行分析。

第 4 章的学习目标:RC 电路和 RL 电路(一阶动态电路)

(1) 熟练掌握电感和电容的特性、VCR 和串并联等效;

(2) 理解换路定则;

(3) 理解一阶电路、零输入响应电路和零状态响应电路的定义;

(4) 熟练掌握 RL 和 RC 电路的零输入响应的分析方法;

(5) 熟练掌握 RL 和 RC 电路的零状态响应的分析方法;

(6) 理解瞬时响应和稳态响应,熟练掌握全响应的计算方法和三要素法。

4.1 电 感

在第 1 章已经学习了电感器,电感器是能够把电能转化为磁能并存储起来的元件,它的 VCR 表示如下:

$$v = L \frac{\mathrm{d}i}{\mathrm{d}t} \tag{4.1}$$

即电感两端的电压等于"流过"它的电流的变化值乘以电感值,且电压电流需要服从关联参考方向。当电流的变化率为 0 时,即直流电流情况下,电压为 0,电感相当于短路。所以,电感具有导通直流的作用。

如果电压为有界值,则电流为

$$i(t) = \frac{1}{L}\int_{-\infty}^{t} v(\zeta)\mathrm{d}\zeta = \frac{1}{L}\int_{-\infty}^{t_0} v(\zeta)\mathrm{d}\zeta + \frac{1}{L}\int_{t_0}^{t} v(\zeta)\mathrm{d}\zeta = i(t_0) + \frac{1}{L}\int_{t_0}^{t} v(\zeta)\mathrm{d}\zeta \tag{4.2}$$

由式(4.2)可知,某时刻电感的电流不仅与该时刻的电压有关,而且还与此时刻以前的所有电压值有关,这说明电感电流具有记忆性。

在电感电压有界的情况下,电感电流不能跳变,具有连续性。这在分析含有电感的电路时是一个非常重要的性质。

如图 4.1(a)所示,当有 N 个电感串联时,根据 KVL,可得

$$v_s = v_1 + v_2 + \cdots + v_N \tag{4.3}$$

根据电感 VCR 关系,可以进一步得到

$$v_s = L_1 \frac{di_1}{dt} + L_2 \frac{di_2}{dt} + \cdots + L_N \frac{di_N}{dt} \tag{4.4}$$

由于所有电感都是串联的,所以电流为

$$i = i_1 = i_2 = \cdots = i_N \tag{4.5}$$

则

$$v_s = (L_1 + L_2 + \cdots L_N) \frac{di}{dt} \tag{4.6}$$

如果令 $v_s = L_{eq} \frac{di}{dt}$,则

$$L_{eq} = \sum_{n=1}^{N} L_n \tag{4.7}$$

即所有电感串联的总电感值等于它们各自电感的和。图 4.1(b) 中的电路可以被视作图 4.1(a) 中电路的等效电路。

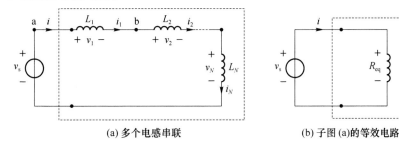

(a) 多个电感串联　　　　(b) 子图 (a) 的等效电路

图 4.1　串联电感

根据 KCL 和电感 VCR,图 4.2 中电路的电流为

$$\begin{aligned} i_s &= i_1 + i_2 + \cdots + i_N = \int \frac{v}{L_1} dt + \int \frac{v}{L_2} dt + \cdots + \int \frac{v}{L_N} dt \\ &= \left(\frac{1}{L_1} + \frac{1}{L_2} + \cdots + \frac{1}{L_N} \right) \int v \, dt \end{aligned} \tag{4.8}$$

如果令 $i_s = \int \frac{v}{L_{eq}} dt$,则

$$\frac{1}{L_{eq}} = \sum_{n=1}^{N} \frac{1}{L_n} \tag{4.9}$$

即并联电路中的总电感的倒数等于各支路电感的倒数和。图 4.2(b) 中的电路可以被视作图 4.2(a) 中电路的等效电路。

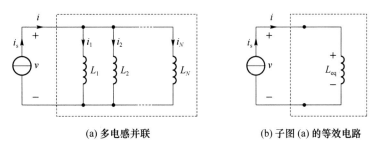

(a) 多电感并联　　　　(b) 子图 (a) 的等效电路

图 4.2　并联电感

4.2 电　　容

在第 1 章已经学习了电容是一种在电场中储存能量的无源双端电路元件,它的 VCR 表示如下:

$$i = C \frac{dv}{dt} \tag{4.10}$$

即为流过电容的电流等于它两端的电压的变化值乘以电容值,且电压电流需要服从关联参考方向。当电压变化率为 0 时,即直流电压情况下,电流为 0,电容相当于开路。因此,电容有隔断直流的作用。

如果电流为有界值,则电压为

$$\begin{aligned} v(t) &= \frac{1}{C}\int_{-\infty}^{t} i(\zeta)d\zeta = \frac{1}{C}\int_{-\infty}^{t_0} i(\zeta)d\zeta + \frac{1}{C}\int_{t_0}^{t} i(\zeta)d\zeta \\ &= v(t_0) + \frac{1}{C}\int_{t_0}^{t} i(\zeta)d\zeta \end{aligned} \tag{4.11}$$

由式(4.11)可知,某时刻的电容电压不仅与该时刻的电流有关,而且还与此时刻以前的所有电流值有关,这说明电容电压具有记忆性。

在电容电流有界的情况下,电容电压不能跳变,具有连续性。这在分析含有电容的电路时是一个非常重要的性质。

如图 4.3(a)所示,当有 N 个电容串联时,根据 KVL,可得

$$v_s = v_1 + v_2 + \cdots + v_N \tag{4.12}$$

根据电容 VCR 关系,可以进一步得到

$$v_s = \int \frac{i_1}{C_1} dt + \int \frac{i_2}{C_2} dt + \cdots + \int \frac{i_N}{C_N} dt \tag{4.13}$$

由于所有电容都是串联的,所以电流为

$$i = i_1 = i_2 = \cdots = i_N \tag{4.14}$$

则

$$v_s = \left(\frac{1}{C_1} + \frac{1}{C_2} + \cdots + \frac{1}{C_N} \right) \int i\, dt \tag{4.15}$$

如果令 $v_s = \frac{1}{C_{eq}} \int i\, dt$,则

$$\frac{1}{C_{eq}} = \sum_{n=1}^{N} \frac{1}{C_n} \tag{4.16}$$

即串联电路中的总电容的倒数等于各支路电容的倒数和。图 4.3(b)中的电路可以被视作图 4.3(a)中电路的等效电路。

根据 KCL 和电容 VCR,图 4.4 中电路的电流为

$$i_s = i_1 + i_2 + \cdots + i_N = C_1 \frac{dv}{dt} + C_2 \frac{dv}{dt} + \cdots + C_N \frac{dv}{dt} \tag{4.17}$$

如果令 $i_s = C_{eq} \frac{dv}{dt}$,则

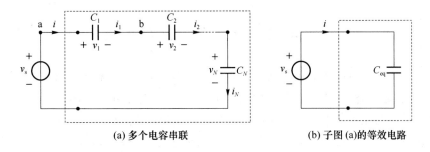

(a) 多个电容串联　　　　　　(b) 子图(a)的等效电路

图 4.3　串联电容

$$C_{eq} = \sum_{n=1}^{N} C_n \tag{4.18}$$

即所有并联电容的总电容值等于它们各自电容的和。图 4.4(b)中的电路可以被视作图 4.4(a)中电路的等效电路。

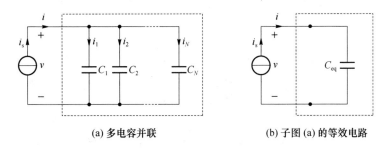

(a) 多电容并联　　　　　　(b) 子图(a)的等效电路

图 4.4　并联电容

4.3　换路定则

电路通常工作在稳定状态,此时电路中的电压和电流为固定值或是发生周期性变化。当电路由一种工作状态变化到另一种工作状态(如电源或开关的连接和断开,电路元件参数的改变等)被称为电路的换路。换路会使电路中的电压和电流发生变化,这种变化会持续一段时间,这个变化过程称为瞬态。

换路之后,可对换路后的电路重新进行分析,在某些情况下其与换路前电路的状态变量存在联系。如果电路中存在电容或电感等记忆性元件,由于电容电压和电感电流的记忆性,在任何时刻都无法发生跳变。因此,在换路时刻 t_0 的前后电容电压与电感电流不能发生跳变,记作

$$v_C(t_0^+) = v_C(t_0^-) \tag{4.19}$$

$$i_L(t_0^+) = i_L(t_0^-) \tag{4.20}$$

其中,$v_C(t_0^-)$ 和 $i_L(t_0^-)$ 为电路换路之前的相应变量的稳定状态。

式(4.19)和式(4.20)就是换路定则。电容电压和电感电流满足换路定则,由此可推出它们换路之后的初始值,并基于此继续推导出其他变量换路之后的初始值。

4.4 一阶电路

之前几章主要针对电源和电阻构成的电路进行分析和讨论,本章开始重点讨论由电源、电阻,以及电感或电容(电容和电感不同时出现)等动态元件组成的电路。

含有一个动态元件的电路被称为一阶电路,通常用一阶微分方程表示。如果原电路中的多个电感或多个电容可以等效为一个电感或一个电容,则该电路可视为一阶电路。常用的一阶电路包括 RL(电阻-电感)电路和 RC(电阻-电容)电路。

分析一阶电路时可以分为以下三种情况。

(1) 当储存在电感或电容中的能量突然释放到电阻网络中时,电路中产生电流和电压。当电感或电容突然与直流电源断开时,会发生这种情况。此时可将电路简化为图 4.5 所示的两种等效形式。这种由动态(储能)元件内部已存储的能量在电路中引起的响应,被称为零输入响应。零输入响应强调:决定电路中响应的是电路本身的性质,而不是外部的激励源。

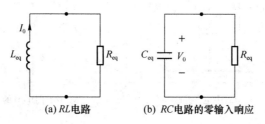

(a) RL 电路 (b) RC 电路的零输入响应

图 4.5 RL 电路和 RC 电路的零输入响应

(2) 当外加直流电压源或电流源时,电感或电容吸收能量,电路产生电流和电压。当电感或电容与直流电源间的开关闭合时,就会发生这种情况。此时可将电路简化为图 4.6 所示的两种等效形式。这种由于电路中储能元件初始状态能量为零,通过外加电源在电路中引起的响应,被称为零状态响应。零状态响应强调:电路中响应来自外部的激励源,而不是动态元件中的初始储能。

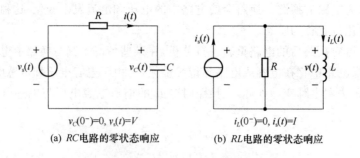

(a) RC 电路的零状态响应 (b) RL 电路的零状态响应

图 4.6 RC 电路和 RL 电路的零状态响应

(3) 电路中既有含有初始储能的动态元件,也有外部激励电源,此时电路的响应称为全响应。根据第 3 章所学的电路叠加定理可知,全响应可以看作是零输入响应和零状态响应的叠加。图 4.7 所示是全响应电路的两种等效形式。

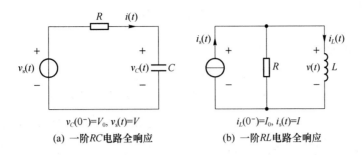

(a) 一阶RC电路全响应   (b) 一阶RL电路全响应

图 4.7 RC 电路和 RL 电路的全响应

4.5 RL 和 RC 电路的零输入响应

本节分析了 RC 电路和 RL 电路的零输入响应。

4.5.1 RC 电路的零输入响应

图 4.8 所示是一个 RC 电路,该电路由直流电压源 V_0,电阻 R_1 和 R,电容 C 和开关 k 组成。$t=0$ 时刻前,开关 k 一直处于位置 a。$t=0$ 时刻,开关 k 从位置 a 移动到位置 b。下面分析 $t>0$ 之后的电容电压 $v_C(t)$。

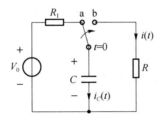

图 4.8 RC 电路的零输入响应

当 $t<0$ 时,开关 k 处于位置 a,电路为电压源串联电阻 R_1 和电容 C。当电源为直流电压源时,电容可视为开路。经过一段时间后,电容两端电压等于直流电压源电压,且电路达到稳定状态。

用 $t=0^-$ 来表示切换前的瞬间,用 $t=0^+$ 表示切换后的初始瞬间。由于电容电压的连续性,不会发生跳变。因此,电容两端的电压在从位置 a 切换到位置 b 后的瞬间是不变的,则切换后电容电压的初始值为

$$v_C(0^+) = v_C(0^-) = V_0 \tag{4.21}$$

当 $t>0$ 时,开关 k 处于位置 b,电路为电阻 R 和电容 C 串联。根据 KVL 方程,可知

$$v_C(t) = i(t)R \tag{4.22}$$

根据电容 VCR

$$i(t) = -i_C(t) = -C\frac{dv_C(t)}{dt} \tag{4.23}$$

由式(4.22)和式(4.23)可得一个一阶齐次微分方程

$$v_C(t) + RC\frac{dv_C(t)}{dt} = 0 \quad (4.24)$$

结合(4.21)所示电容电压的初始值,该方程的解为

$$v_C(t) = V_0 e^{-\frac{t}{RC}} \quad (4.25)$$

其中,V_0 为电容电压换路之后的初始值。此外,电容电压的变化还与 RC 值的大小有关。

此处,引入一个新概念——时间常数。在式(4.25)中,$\frac{1}{RC}$ 确定电流或电压趋近于零的速率。这个速率的倒数定义为电路的时间常数 τ,时间常数等于电阻与电容的乘积,可记作

$$\tau = RC \quad (4.26)$$

由式(4.26)可知,在零输入响应中时间常数 RC 控制电流或电压的衰减速率。图 4.9 描绘了图 4.8 所示 RC 电路的电容电压的零输入响应曲线。

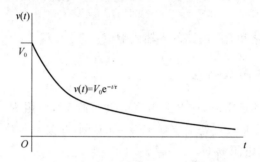

图 4.9 RC 电路的零输入响应曲线

因此,计算 RC 电路电容电压的零输入响应可分为以下几个步骤:

(1) 计算电容电压换路后的初始值 $v_C(0^+) = v_C(0^-)$;
(2) 写出电路的 KVL 方程和电容的 VCR 方程构成一阶齐次微分方程;
(3) 计算时间常数 $\tau = RC$;
(4) 求解方程,得到换路后电容电压 $v_C(t) = v_C(0^+) e^{-\frac{t}{\tau}}$。

4.5.2 RL 电路的零输入响应

图 4.10 所示是一个 RL 电路,该电路由直流电流源 i_0,电阻 R_0 和 R,电感 L 和开关 k 组成。$t=0$ 时刻前,开关 k 一直处于闭合状态。$t=0$ 时刻,开关打开。下面分析 $t>0$ 之后的电感电流 $i_L(t)$。

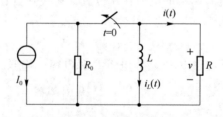

图 4.10 RL 电路零输入响应

当 $t<0$ 时,开关闭合,电路为电流源并联电阻 R_0、R 和电感 L。当电源为直流电流源

时,电感可视为短路。经过一段时间后,流过电感的电流等于直流电流源电流,且电路达到稳定状态。

用 $t=0^-$ 来表示开关打开前的瞬间,用 $t=0^+$ 表示开关打开后的初始瞬间。由于电感电流的连续性,不会发生跳变。因此,电感电流在开关打开瞬间是不变的,开关打开后电感电流的初始值为

$$i_L(0^+)=i_L(0^-)=I_0 \tag{4.27}$$

当 $t>0$ 时,开关 k 打开,电路为电阻 R 和电感 L 串联。根据 KVL 方程,可知

$$v_L(t)=i(t)R \tag{4.28}$$

根据电感 VCR

$$v_L(t)=L\frac{\mathrm{d}i_L}{\mathrm{d}t}=-L\frac{\mathrm{d}i}{\mathrm{d}t} \tag{4.29}$$

由式(4.28)和式(4.29)可得一个一阶齐次微分方程

$$i_L(t)+\frac{L}{R}\frac{\mathrm{d}i_L(t)}{\mathrm{d}t}=0 \tag{4.30}$$

结合式(4.27)所示电感电流的初始值,该方程的解为

$$i_L(t)=I_0\mathrm{e}^{-\frac{Rt}{L}} \tag{4.31}$$

其中,I_0 为电感电流换路之后的初始值。此外,电感电流的变化还与 $\frac{L}{R}$ 值的大小有关。

值得注意的是,电感电流的表达式包含一项 $\mathrm{e}^{-\frac{Rt}{L}}$,其中 $\frac{R}{L}$ 决定了电流或电压趋近于零的速率。这个比值的倒数是该电路的时间常数 $\tau=\frac{L}{R}$。

由式(4.31)可知,在零输入响应中时间常数 $\tau=\frac{L}{R}$ 控制电流或电压的衰减速率。图 4.11 所示,描绘了图 4.10 所示 RL 电路的电感电流的零输入响应曲线。

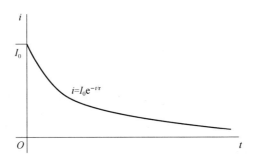

图 4.11 RL 电路的零输入响应曲线

如前所述,RL 电路的零输入响应类似于 RC 电路。RL 电路电感电流的零输入响应计算可分为以下几个步骤:

(1) 计算电感电流换路后的初始值 $i_L(0^+)=i_L(0^-)=I_0$;

(2) 写出电路的 KVL 方程和电感的 VCR 方程构成一阶齐次微分方程;

(3) 计算时间常数 $\tau=\frac{L}{R}$;

(4) 求解方程,得到换路后的电感电流 $i_L(t)=i_L(0^+)\mathrm{e}^{-\frac{t}{\tau}}$。

4.6 RL 和 RC 电路的零状态响应

当外加直流电压源或电流源时,电感或电容吸收能量,电路产生电流和电压。这种由于电路中储能元件初始状态能量为零,通过外加电源在电路中引起的响应,被称为零状态响应。零状态响应强调:电路中响应来自外部的激励源,而不是动态元件中的初始储能。

如图 4.12 所示,阶跃函数可以用来描述直流电源。例如,阶跃函数 $v(t)$ 可以表示为

$$v(t)=\begin{cases}V_0, & t\geqslant t_0 \\ 0, & t<t_0\end{cases} \quad (4.32)$$

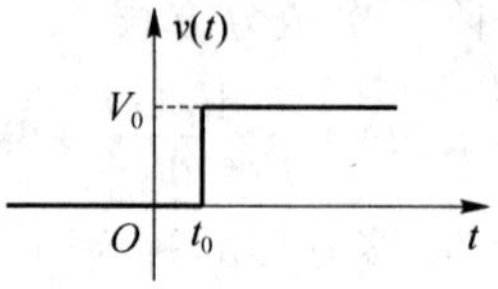

图 4.12 阶跃函数

本节分析了 RC 电路和 RL 电路的零状态响应。

4.6.1 RL 电路的零状态响应

图 4.13 所示是一个 RL 电路,该电路由直流电压源 $v_s(t)\big|_{t\geqslant 0}=V_s$,电阻 R,电容 L 和开关 k 组成。开关 k 在 $t=0$ 时刻关闭。下面分析 $t>0$ 后的电感电流 $i_L(t)$。

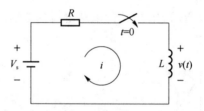

图 4.13 RL 电路零状态响应的说明

在开关 k 闭合前,RL 电路中的电流为零,即 $i_L(0^-)=0$。当开关 k 关闭后,根据电路 KVL 方程和电感 VCR 得到一个非齐次微分方程,记作

$$Ri(t)+L\frac{\mathrm{d}i(t)}{\mathrm{d}t}=v_s(t) \quad (4.33)$$

求解非齐次微分方程的步骤如下:

当初始条件为 $y(0)=Y_0$ 时,微分方程 $a_1y'(t)+a_0y(t)=f(t)$ 的解为 $y(t)=y_h(t)+y_p(t)$,其中,$y_h(t)$ 为齐次方程 $a_1y'(t)+a_0y(t)=0$ 的解,$y_p(t)$ 为方程 $a_0y(t)=f(t)$ 的特解。

齐次方程 $Ri(t)+L\dfrac{\mathrm{d}i(t)}{\mathrm{d}t}=0$ 的解为

$$i_h(t)=A\mathrm{e}^{-\frac{t}{\tau}} \quad (4.34)$$

其中，$\tau = \dfrac{L}{R}$ 且 A 是一个待确定的系数。

特解为

$$i_p(t) = \dfrac{V_s}{R} \tag{4.35}$$

因此结果为 $i(t) = i_h(t) + i_p(t) = A\mathrm{e}^{-\frac{t}{\tau}} + \dfrac{V_s}{R}$。

结合初始条件，得到 $i(0) = A + \dfrac{V_s}{R} = 0$，即 $A = -\dfrac{V_s}{R}$。因此，图 4.13 中 RL 电路的零状态响应为

$$i(t) = \dfrac{V_s}{R}(1 - \mathrm{e}^{-\frac{t}{\tau}}) \tag{4.36}$$

式(4.36)中的 $\dfrac{V_s}{R}$ 为电感电流的稳定值 $i_L(\infty) = \dfrac{V_s}{R}$，时间常数 $\tau = \dfrac{L}{R}$ 控制衰减速率。图 4.14 描绘了 RL 电路的零状态响应。

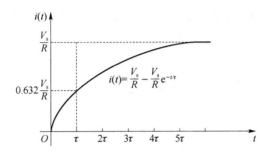

图 4.14　RL 电路的零状态响应

4.6.2　RC 电路的零状态响应

图 4.15 所示是一个 RC 电路。该电路由直流电流源 $i_s(t)\big|_{t \geqslant 0} = I_s$、电阻 R、电容 C 和开关 k 组成。开关 k 在 $t = 0$ 时刻关闭。下面分析 $t > 0$ 后的电容电流 $v_C(t)$。

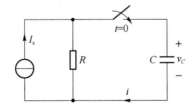

图 4.15　RC 电路零状态响应的说明

在开关 k 闭合前，RC 电路中的电容两端电压为零，即 $v_C(0^-) = 0$。当开关 k 关闭后，根据电路 KCL 方程和电容 VCR 得到一个非齐次微分方程，记作

$$C\dfrac{\mathrm{d}v_C(t)}{\mathrm{d}t} + \dfrac{v_C(t)}{R} = i_s(t), \quad t \geqslant 0 \tag{4.37}$$

结合初始条件 $v_C(0) = 0$，得到非齐次方程的解

$$v_C(t) = RI_s(1 - e^{-\frac{t}{RC}}) \qquad (4.38)$$

式(4.38)中的 RI_s 为电容电压的稳定值 $v_C(\infty) = RI_s$。

综上所述，一阶电路的零状态响应计算可分为三个步骤：

(1) 计算电路变量换路后的稳定值 $v_C(\infty)$ 或 $i_C(\infty)$；

(2) 计算时间常数 τ；

(3) 得到换路后的变量 $v_C(t) = v_C(\infty)(1 - e^{-\frac{t}{\tau}})$ 或 $i_C(t) = i_C(\infty)(1 - e^{-\frac{t}{\tau}})$。

4.7 RL 电路和 RC 电路的全响应和三要素法

前面的两小节分别介绍了一阶电路的零输入响应和零状态响应。通常情况下，电路中的动态元件既有初始储能，电路又有外加激励电源，这种情况下的电路响应为全响应。对于线性非时变动态电路，根据电路的叠加定理，电路的全响应等于电路的零输入响应和零状态响应之和。

如图 4.16 所示，开关 k 闭合前电容已有初始储能，$v_C(0^-) = V_0$，直流电压源 $v_s(t)\big|_{t \geq 0} = V_s$，则开关闭合后电路的响应为全响应。

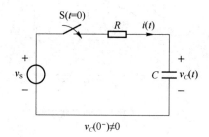

图 4.16 RC 电路

根据电路的 KVL 方程和元件的 VCR 关系，可得

$$RC\frac{dv_C(t)}{dt} + V_C(t) = V_s \qquad (4.39)$$

通过求解非齐次微分方程，可得特解 $v_{cp}(t) = V_s$，通解为指数形式 $v_{ch}(t) = Ae^{-\frac{1}{RC}t}$。

根据电容电压初始值可求得系数

$$A = v_C(0^+) - V_s = V_0 - V_s \qquad (4.40)$$

所以电容电压的全响应为

$$v_C(t) = v_{ch}(t) + v_{cp}(t) = V_s + (V_0 - V_s)e^{-\frac{1}{RC}t} \qquad (4.41)$$

重写式(4.41)可得

$$v_C(t) = V_0 e^{-\frac{1}{RC}t} + V_s(1 - e^{-\frac{1}{RC}t}) \qquad (4.42)$$

其中，$V_0 e^{-\frac{1}{RC}t}$ 是电容电压的零输入响应，$V_s(1 - e^{-\frac{1}{RC}t})$ 是电容电压的零状态响应。式(4.42)验证了电路的全响应等于电路的零输入响应和零状态响应之和。

通过上述分析可知，在直流电源激励下，动态电路中任一支路的电压和电流都可以利用下式直接求解

$$y(t)=y(\infty)+[y(t_0^+)-y(\infty)]e^{-\frac{t-t_0}{\tau}} \tag{4.43}$$

其中，$y(t_0^+)$ 为初始值，$y(\infty)$ 为稳态值，t_0 是换路时刻，τ 是时间常数。

无论是求解零输入响应、零状态响应还是全响应，只要知道了待求变量的初始值、稳态值和时间常数，就可以直接写出响应结果，这种方法被称为三要素法。表 4.1 总结了一阶电路的零输入响应、零状态响应和全响应的三要素求解方法。

表 4.1 一阶电路响应总结

响应类型	电路类型	解的形式（t_0 是换路时刻）	时间常数
零输入响应	RC 电路	$y(t)=y(t_0^+)e^{-\frac{t-t_0}{\tau}}$ $y(t)$ 为 $v_C(t)$ 或 $i_C(t)$	$\tau=RC$
	RL 电路		$\tau=\dfrac{L}{R}$
零状态响应	RC 电路	$y(t)=y(\infty)\left(1-e^{-\frac{t-t_0}{\tau}}\right)$ $y(t)$ 为 $v_C(t)$ 或 $i_C(t)$	$\tau=RC$
	RL 电路		$\tau=\dfrac{L}{R}$
全响应	RC 电路	$y(t)=y(\infty)+[y(t_0^+)-y(\infty)]e^{-\frac{t-t_0}{\tau}}$	$\tau=RC$
	RL 电路		$\tau=\dfrac{L}{R}$

问题 4.1 图 4.17 所示的电路中，$t<0$ 时开关一直处在位置 1。在 $t=0$ 时，开关 S 从位置 1 移动到位置 2。计算 $t>0$ 后的电容电流 i_C。

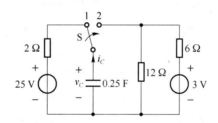

图 4.17 问题 4.1 中的电路

解： 电容两端的初始电压为

$$v_C(0^+)=v_C(0^-)=25 \text{ V} \tag{4.44}$$

开关切换后，使用戴维南等效定理对电容右边的电路进行分析。它的开路电压为

$$v_{\text{oc}}=\frac{12 \text{ Ω} \times 3 \text{ V}}{12 \text{ Ω}+6 \text{ Ω}}=2 \text{ V} \tag{4.45}$$

等效电阻为

$$R_{\text{eq}}=12 \text{ Ω} // 6 \text{ Ω}=4 \text{ Ω} \tag{4.46}$$

该电路可以等效为如图 4.18 所示。

因此，稳态电压为

$$v_C(\infty)=2 \text{ V} \tag{4.47}$$

时间常数为

$$\tau=R_{\text{eq}}C=4 \text{ Ω} \times 0.25 \text{ F}=1 \text{ s} \tag{4.48}$$

则电路的全响应为

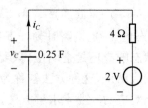

图 4.18 开关状态改变后的等效电路

$$v_C(t) = v_C(\infty) + [v_C(0^+) - v_C(\infty)]e^{-t/\tau} \quad (4.49)$$
$$= 2 + (25-2)e^{-t} = 2 + 23e^{-t} \text{ V}$$

因此,流过电容的电流为

$$i_C(t) = -C\frac{dv_C(t)}{dt} = -0.25 \times (-23)e^{-t} = 5.75e^{-t} \text{ A} \quad (4.50)$$

4.8 习 题

习题 4.1 图 4.19 中的开关在 $t=0$ 时刻打开之前,一直是处在闭合的状态。计算 $t>0$ 后的电流 i。

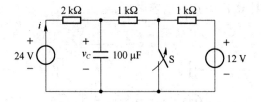

图 4.19 习题 4.1 中的电路

习题 4.2 图 4.20 所示的电路中的开关在 $t<0$ 一直处在位置 1。在 $t=0$ 时,开关 S 从位置 1 移动到位置 2。计算 $t>0$ 后的电压 $v(t)$。

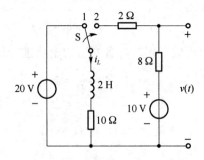

图 4.20 习题 4.2 中的电路

习题 4.3　图 4.21 中的开关在 $t=0$ 时刻关闭之前,一直处在打开的状态。计算 $t>0$ 时后的电压 v_L。

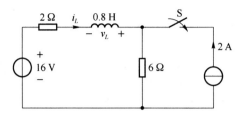

图 4.21　习题 4.3 中的电路

习题 4.4　图 4.22 中的开关在 $t=0$ 时刻关闭之前,一直处在打开的状态。计算 $t>0$ 之后的电流 i_1。

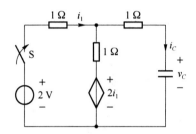

图 4.22　习题 4.4 中的电路

习题 4.5　图 4.23 中的开关在 $t=0$ 时刻关闭之前,一直处在打开的状态。计算 $t>0$ 之后的电压 v_L。

（a）计算 $i_1(0^-)$、$i_2(0^-)$、$i_1(0^+)$ 和 $i_2(0^+)$；
（b）计算 $t>0$ 后的 $i_1(t)$ 和 $i_2(t)$；
（c）解释为什么 $i_2(0^-)\neq i_2(0^+)$。

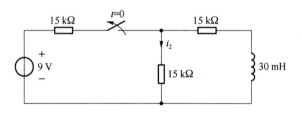

图 4.23　习题 4.5 中的电路

习题 4.6　图 4.24 中的开关在 $t=0$ 时刻之前,一直处在闭合的状态。计算 $t>0$ 之后的电压 $v_o(t)$。

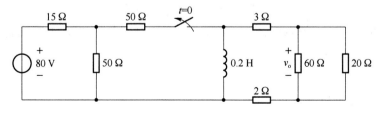

图 4.24　习题 4.6 中的电路

习题 4.7　图 4.25 中的开关在 $t=0$ 时刻之前，一直处在打开的状态。计算 $t>0$ 之后的电压 v_L。

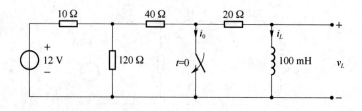

图 4.25　习题 4.7 中的电路

习题 4.8　图 4.26 中的开关之前一直处在 a 位置。在 $t=0$ 时刻，它立刻从 a 位置移动到 b 位置。

(a) 计算 $t>0$ 之后的 $v_o(t)$ 的值；

(b) 1 kΩ 电阻吸收的能量为多少？

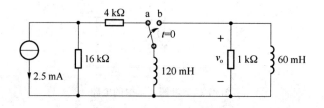

图 4.26　习题 4.8 中的电路

习题 4.9　图 4.27 中的开关之前一直处在 a 位置。在 $t=0$ 时刻，它立刻从 a 位置移动到 b 位置。接下来计算：

(a) 假设 $v_2(0^-)=0$，计算 $t \geqslant 0$ 之后的 i、v_1 和 v_2；

(b) 计算 $t=0$ 时刻，电容中储存的能量；

(c) 如果开关保持在 b 位置，计算电路中获得的电能和 5 kΩ 电阻中消耗的总电能。

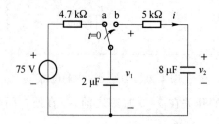

图 4.27　习题 4.9 中的电路

习题 4.10　图 4.28 所示电路，$v_{C1}(0^-)=v_{C2}(0^-)=0$，$t=0$ 时刻开关闭合，求 $i_{C1}(0^+)$，$i_{C2}(0^+)$，$v_{C1}(\infty)$ 和 $v_{C2}(\infty)$。

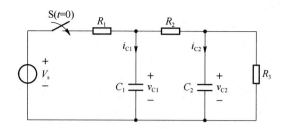

图 4.28 习题 4.10 中的电路

习题 4.11 题图 4.29 所示电路在开关闭合前已达稳态，$t=0$ 时刻开关闭合，求开关闭合后的电容电压 $v_C(t)$。

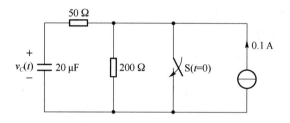

图 4.29 习题 4.11 中的电路

习题 4.12 题图 4.30 所示电路在开关打开前已达稳态，$t=0$ 时刻开关打开，求开关打开后的电容电压 $v_C(t)$ 和电流 $i(t)$。

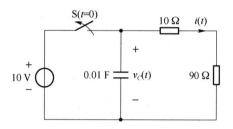

图 4.30 习题 4.12 中的电路

习题 4.13 题图 4.31 所示电路，开关在位置 1 闭合已久，$t=0$ 时刻开关由 1 合向 2，求 $t \geqslant 0^+$ 后的电感电流 i_L。

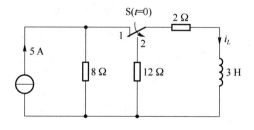

图 4.31 习题 4.13 中的电路

习题 4.14 题图 4.32 所示电路在 $t=0^-$ 时已处于稳态，$t=0$ 时刻开关 S_1 闭合，S_2 打开。

(a) 求 $t \geqslant 0$ 后的电感电流 i_L；

(b) 若电路中电源电压改为 60 V，求 $t \geqslant 0$ 后的电感电流 i_L。

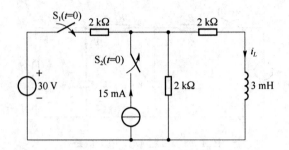

图 4.32 习题 4.14 中的电路

习题 4.15 求题图 4.33 所示电路在下列条件下 $t \geqslant 0$ 时的 $i_L(t)$。
(a) $v_s(t)=0\,\text{V}, i_L(0^-)=3\,\text{A}$；
(b) $v_s(t)=10\,\text{V}, i_L(0^-)=0$；
(c) $v_s(t)=20\,\text{V}, i_L(0^-)=-1\,\text{A}$。

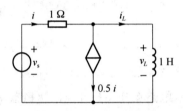

图 4.33 习题 4.15 中的电路

习题 4.16 电路如题图 4.34 所示，开关在 $t=0$ 时闭合。已知 $v_C(0^-)=1\,\text{V}, v_s=1\,\text{V}$。求 $t \geqslant 0$ 时，该电路中 $2\,\Omega$ 电阻上的电压 $v_0(t)$。

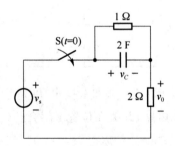

图 4.34 习题 4.16 中的电路

习题 4.17 已知 $t<0$ 时题图 4.35 所示电路已处于稳态，求 $t \geqslant 0$ 后的 $v_{ab}(t)$。

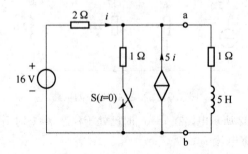

图 4.35 习题 4.17 中的电路

习题4.18 题图4.36所示电路，$v_C(0^-)=2\text{V}$，求 $t\geqslant 0^+$ 后的 $v_C(t)$。

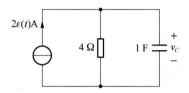

图4.36 习题4.18中的电路

习题4.19 用单位阶跃函数表示题图4.37所示的3个电压波形。

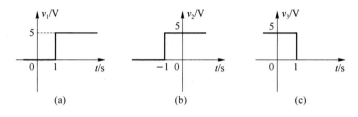

图4.37 习题4.19中的电路

习题4.20 图4.38给出了电路结构以及 v_{in} 的波形。假设 $T=5RC$，求 $t>0$ 之后的 v_o。

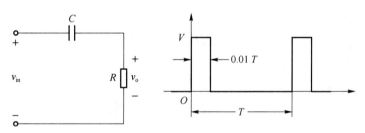

图4.38 习题4.20中的电路

第 5 章

正弦稳态电路分析

到目前为止,本书分析的所有电路的共同特点是电源为直流电源。在本章将带领大家一起学习含有时变电压源或电流源(交流电源)的电路。特别是分析电压源或电流源的变化是正弦变化的情况。第 2~4 章已经学习的电路分析方法和定理在时变电源电路中仍然成立。因此,本章中的一些分析方法和结论大家看起来会比较熟悉。正弦分析的难点主要是在复数域上进行数学分析。

第 5 章的学习目标:正弦稳态电路分析:

(1) 理解正弦信号、正弦电源和正弦响应的基本概念;
(2) 掌握正弦变量的相量表示方法;
(3) 理解频域内阻抗和导纳的概念;
(4) 理解并掌握频域上的电路基本分析方法和电路基本定理;
(5) 理解复数功率,并掌握复数功率计算方法。

5.1 正弦函数

如图 5.1 所示,正弦函数是随时间呈正弦变化的函数,由幅值、角频率和初始相位确定。

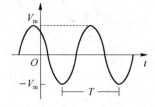

图 5.1 正弦函数

本章用余弦函数作为正弦曲线函数,记作

$$v(t) = V_m \cos(\omega t + \phi) \tag{5.1}$$

其中:

(1) ϕ 为 $t=0$ 时的初始相位。若 $\phi=0$,函数如图 5.2(a)所示;若 $\phi<0$,函数如图 5.2(b)所示;若 $\phi>0$,函数如图 5.2(c)所示。

(2) ω 为角频率,其计算公式为 $\omega = 2\pi f = \dfrac{2\pi}{T}$,其中频率 f 的单位是赫兹(Hz),周期 T 的单位是秒(s)。

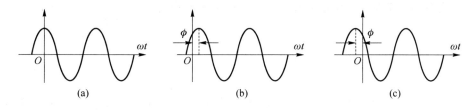

图 5.2 正弦函数

(3) V_m 是最大幅度值。

正弦函数的另一个重要特征是均方根。它是定义正弦电源电压有效值或电流有效值的一种最普遍的数学方法。周期函数的均方根定义为函数平方均值的平方根。因此，$v(t)=V_m\cos(\omega t+\phi)$ 的均方根为

$$V_{rms}=\sqrt{\frac{1}{T}\int_{t_0}^{t_0+T}v(t)^2dt}=\sqrt{\frac{1}{T}\int_{t_0}^{t_0+T}V_m^2\cos^2(\omega t+\phi)dt}=\frac{V_m}{\sqrt{2}} \tag{5.2}$$

正弦函数的均方根只取决于最大振幅 V_m，与频率或相位角无关。后面将强调均方根的重要性，因为它与功率计算有关。

5.2 正弦响应

图 5.3 所示为一种常见的简单电路。

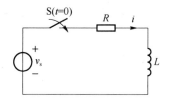

图 5.3 一种常见的简单电路

电压源提供的电压为

$$v_s=V_m\cos(\omega t+\phi) \tag{5.3}$$

不同于第 2 章中的直流电源，这里的电压源是一个时变的正弦电压。根据 KVL 方程和元件的 VCR 关系，对图 5.3 所示的电路可写出

$$L\frac{di_L}{dt}+Ri_L=V_m\cos(\omega t+\phi) \tag{5.4}$$

式(5.4)的解为

$$i_L(t)=\frac{-V_m}{\sqrt{R^2+\omega^2L^2}}\cos(\phi-\theta)e^{-(R/L)t}+\frac{V_m}{\sqrt{R^2+\omega^2L^2}}\cos(\omega t+\phi-\theta) \tag{5.5}$$

其中，$\theta=\arctan(\omega L/R)$。式(5.5)右边的第一项称为电流的瞬时分量，因为其随着时间的推移会变得无穷小。右边的第二项称为稳态分量，对于稳态分量，只要电源提供电压，稳态响应就存在。由于瞬时响应随着时间的推移而消失，稳态响应单独也满足(5.5)所示的微分方程。在线性电路中，如果电源为正弦函数，稳态响应也是一个正弦函数，且稳态响应的频率

与正弦电源的频率相同,但振幅和相位与正弦电源通常不同。

5.3 相　　量

由于同一电路中的各变量频率相同,可以仅用电路变量的最大值(幅值)和初始相位来表示变量,简化表示方式。因此,可以把正弦函数转换成一个由幅值和初始相位表示的复数。

相量的概念源于欧拉公式,是表示正弦函数的幅值和相位的一个复数,将指数函数与三角函数联系起来

$$e^{j\phi} = \cos\phi + j\sin\phi \tag{5.6}$$

余弦函数可以看作是指数函数的实部,正弦函数是指数函数的虚部,即 $\cos\phi = \text{Re}\{e^{j\phi}\}$ 和 $\sin\phi = \text{Im}\{e^{j\phi}\}$,其中 $\text{Re}\{g\}$ 表示取实部,$\text{Im}\{g\}$ 表示取虚部。

5.3.1 相量变换

根据欧拉公式,正弦函数可以表示为

$$v(t) = V_m\cos(\omega t + \phi) = \text{Re}\{V_m e^{j(\omega t + \phi)}\} = \text{Re}\{V_m e^{j\phi} e^{j\omega t}\} \tag{5.7}$$

式(5.7)最后一项中的 $V_m e^{j\phi}$ 是一个复数,它包含正弦函数的幅值和相位信息,而 $e^{j\omega t}$ 仅表示了电路的频率信息。因此,在变量的表示中,我们只关心 $V_m e^{j\phi}$ 部分包含的信息。根据定义,复数是正弦函数的相量变换或相量表示。因此,$V_m\cos(\omega t + \phi)$ 的相量变换或相量表示记作 $\dot{V}$ 或 $\mathscr{P}\{V_m\cos(\omega t + \phi)\}$,且

$$\mathscr{P}\{V_m\cos(\omega t + \phi)\} = V_m e^{j\phi} \tag{5.8.a}$$

或

$$\dot{V} = V_m e^{j\phi} \tag{5.8.b}$$

因此,相量变换将正弦函数从时域映射到复数域。相量还可以用以下几种形式表示

$$\dot{V} = V_m e^{j\phi} = V_m \angle \phi = V_m(\cos\phi + j\sin\phi) \tag{5.9}$$

相量变换有助于简化正弦电源下的电路分析,可以以简单明了的形式求解交流电路中稳态正弦响应的幅值和相位。

5.3.2 相量反变换

5.3.1 小节介绍的相量变换过程是可逆的,其逆变换是相量反变换。也就是说,可以把相量从复数域上转换成时域上的正弦函数。相量 $V_m e^{j\phi}$ 的相量反变换可以表示为 $\mathscr{P}^{-1}\{V_m e^{j\phi}\}$,且

$$\mathscr{P}^{-1}\{V_m e^{j\phi}\} = \text{Re}\{V_m e^{j\phi} e^{j\omega t}\} = V_m\cos(\omega t + \phi) \tag{5.10}$$

注意:ω 的值无法由相量推导出来,但其与电路中电源的频率相同。

综上所述,相量变换和相量反变换可以在时域和复数域之间来回转换。

5.3.3 相量的基本运算

本节介绍相量之间的集中基本运算法则,具体包括相量的加(减)法、微分和积分运算。

1. 加减运算

如果正弦函数表示为 $v_n(t) = V_{mn}\cos(\omega t + \phi_n)$,那么该函数求和的相量表示为 $\mathscr{P}\left\{\sum_{n=1}^{N} v_n(t)\right\} = \sum_{n=1}^{N} \mathscr{P}\{v_n(t)\}$。

证明:

$$\begin{aligned}
\sum_{n=1}^{N} v_n(t) &= \sum_{n=1}^{N} V_{mn}\cos(\omega t + \phi_n) \\
&= \sum_{n=1}^{N} \text{Re}\{V_{mn}e^{j\phi_n}e^{j\omega t}\} = \text{Re}\left\{\left(\sum_{n=1}^{N} V_{mn}e^{j\phi_n}\right)e^{j\omega t}\right\} \\
&= \text{Re}\left\{\sum_{n=1}^{N} \mathscr{P}\{v_n(t)\}e^{j\omega t}\right\}
\end{aligned} \quad (5.11)$$

因此可得

$$\mathscr{P}\left\{\sum_{n=1}^{N} v_n(t)\right\} = \sum_{n=1}^{N} \mathscr{P}\{v_n(t)\} \quad (5.12)$$

问题 5.1 假设 $y_1(t) = 20\cos(\omega t + 30°)$ 和 $y_2(t) = 40\cos(\omega t + 60°)$,(1)将 $y(t) = y_1(t) + y_2(t)$ 表示为一个正弦函数,利用三角恒等式求解;(2)利用相量变换求解 $y(t)$。

解:(1)使用三角法恒等式可以得到

$$y(t) = 44.72\cos(\omega t + 33.43°) \quad (5.13)$$

(2) 使用相量的加法运算可得

$$\begin{aligned}
\dot{Y} &= \dot{Y}_1 + \dot{Y}_2 \\
&= 20\angle -30° + 40\angle 60° \\
&= (17.32 - j10) + (20 + j34.64) \\
&= 37.32 + j24.64 \\
&= 44.72\angle 33.43°
\end{aligned} \quad (5.14)$$

根据相量反变换可得

$$y(t) = 44.72\cos(\omega t + 33.43°) \quad (5.15)$$

相量的减法运算与加法运算类似,大家可自行证明。

2. 微分运算

如果正弦函数表示为 $v_n(t) = V_{mn}\cos(\omega t + \phi_n)$,那么该函数求微分的相量表示为 $\mathscr{P}\left\{\dfrac{dv(t)}{dt}\right\} = j\omega\mathscr{P}\{v(t)\}$。

证明:

$$\begin{aligned}
\frac{dv(t)}{dt} &= \frac{d(V_m\cos(\omega t + \phi))}{dt} = \frac{d(\text{Re}\{V_m e^{j\phi}e^{j\omega t}\})}{dt} \\
&= \text{Re}\left\{\frac{d(V_m e^{j\phi}e^{j\omega t})}{dt}\right\} = \text{Re}\{V_m e^{j\phi}(j\omega e^{j\omega t})\} = \text{Re}\{j\omega \dot{V}e^{j\omega t}\}
\end{aligned} \quad (5.16)$$

因此可得

$$\mathscr{P}\left\{\frac{\mathrm{d}v(t)}{\mathrm{d}t}\right\}=\mathrm{j}\omega\mathscr{P}\{v(t)\} \tag{5.17}$$

3. 积分运算

如果正弦函数表示为 $v_n(t)=V_{mn}\cos(\omega t+\phi_n)$，那么该函数求积分的相量表示为 $\mathscr{P}\left\{\int v(t)\mathrm{d}t\right\}=\frac{1}{\mathrm{j}\omega}\mathscr{P}\{v(t)\}$。

证明：

$$\int v(t)\mathrm{d}t = \int V_m\cos(\omega t+\phi)\mathrm{d}t = \int\mathrm{Re}\{V_m\mathrm{e}^{\mathrm{j}\phi}\mathrm{e}^{\mathrm{j}\omega t}\}\mathrm{d}t$$

$$=\mathrm{Re}\left\{\int V_m\mathrm{e}^{\mathrm{j}\phi}\mathrm{e}^{\mathrm{j}\omega t}\mathrm{d}t\right\}=\mathrm{Re}\left\{V_m\mathrm{e}^{\mathrm{j}\phi}\cdot\frac{1}{\mathrm{j}\omega}\mathrm{e}^{\mathrm{j}\omega t}\right\}=\mathrm{Re}\left\{\frac{1}{\mathrm{j}\omega}\dot{V}\mathrm{e}^{\mathrm{j}\omega t}\right\} \tag{5.18}$$

因此可得

$$\mathscr{P}\left\{\int v(t)\mathrm{d}t\right\}=\frac{1}{\mathrm{j}\omega}\mathscr{P}\{v(t)\} \tag{5.19}$$

5.4 阻抗和导纳

由动态元件的 VCR 关系可知，其在时域上是微分关系，无法像电阻一样写成简单的线性形式，但是在复数域上其可有与电阻类似的线性表达形式。

电路元件（如电阻、电感、电容）VCR 的相量形式均可以写作如下形式：

$$Z=\frac{\dot{V}}{\dot{I}} \tag{5.20}$$

该式与欧姆定律一样表达的是电压电流的线性关系，因此被称为欧姆定律的相量形式。式(5.20)中电压相量与电流相量之比可以是复数。其中 Z 为元件的阻抗，单位为欧姆。

因为阻抗可能为复数，所以其可以记作如下形式：

$$Z=R+\mathrm{j}X \tag{5.21}$$

其中，R 为实部，X 为虚部被称为电抗。

电路元件（如电阻、电感、电容）的阻抗可分别记作 $Z_R=R$，$Z_L=\mathrm{j}\omega L$，$Z_C=\frac{1}{\mathrm{j}\omega C}=-\mathrm{j}\frac{1}{\omega C}$。其中电感和电容的阻抗均只含有虚部。对电感而言，$X_L=\omega L$ 为电感的电抗，简称感抗；对电容而言，$X_C=-\frac{1}{\omega C}$ 为电容的电抗，简称为容抗。

频率 ω 越小时，X_L 越小，X_C 越大。当 $\omega=0$（直流）时，$X_L=0$，$X_C\to\infty$。这说明电感对于直流无阻碍作用，相当于短路；电容对于直流相当于开路。

电阻、电感、电容元件的导纳可以记作如下形式：

$$Y=\frac{1}{Z}=\frac{\dot{I}}{\dot{V}} \tag{5.22}$$

导纳是阻抗的倒数，单位是西门子(S)，可以衡量电路或器件允许电流流动的容易程度。

导纳和阻抗一样，是一个由实部（电导 G）和虚部（电纳 B）组成的复数，因此可以写成

$$Y=G+\mathrm{j}B \tag{5.23}$$

电路元件（如电阻、电感、电容）的导纳可分别记作 $Y_R=G=\dfrac{1}{R}$, $Y_L=\dfrac{1}{j\omega L}=-j\dfrac{1}{\omega L}$, $Y_C=j\omega C$。其中电感和电容的导纳均只含有虚部。对电感而言，$B_L=-\dfrac{1}{\omega L}$ 为电感的电纳，简称感纳；对电容而言，$B_C=\omega C$ 为电容的电纳，简称为容纳。

复数域电路元件的总结如表 5.1 所示。

表 5.1 复数域电路元件的总结

元件名称	阻抗	电抗	导纳	电纳
电阻	R	—	G	—
电感	$j\omega L$	ωL	$j(-1/\omega L)$	$-1/\omega L$
电容	$j(-1/\omega C)$	$-1/\omega C$	$j\omega C$	ωC

5.5 电路元件 VCR 的相量表示

5.5.1 电阻 VCR 的相量形式

根据欧姆定律，如果电阻中的电流 $i(t)=I_m\cos(\omega t+\phi)$ 随时间呈正弦性变化，则电阻两端的电压为

$$v(t)=RI_m\cos(\omega t+\phi) \tag{5.24}$$

将该电压转换为相量的形式

$$\dot{V}=RI_m e^{j\phi}=RI_m\angle\phi \tag{5.25}$$

因此，电阻 VCR 的相量形式为

$$\dot{V}=R\cdot\dot{I} \tag{5.26}$$

图 5.4 所示为电阻两端的电压和电流随时间的变化，不难发现电阻的电压和电流处于同一相位。

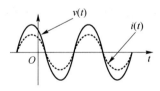

图 5.4 电阻两端的电压和电流随时间的变化

5.5.2 电容 VCR 的相量形式

如果电容两端的电压为 $v(t)=V_m\cos(\omega t+\phi)$，则电容的 VCR 为

$$i(t)=C\dfrac{dv}{dt}=\omega C V_m\cos(\omega t+\phi+90°) \tag{5.27}$$

因此，电容 VCR 的相量形式为

$$\dot{V} = \frac{1}{j\omega C}\dot{I} \tag{5.28}$$

图 5.5 所示为电容两端的电压和电流随时间的变化，不难发现电容的电流超前电压 90°。

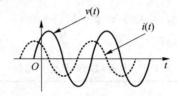

图 5.5　电容两端的电压和电流随时间的变化

5.5.3　电感 VCR 的相量形式

如果电感中的电流为 $i(t)=I_m\cos(\omega t+\phi)$，则电感的 VCR 为

$$v(t)=L\frac{\mathrm{d}i}{\mathrm{d}t}=\omega L I_m\cos(\omega t+\phi+90°) \tag{5.29}$$

因此，电感 VCR 的相量形式为

$$\dot{V}=j\omega L\dot{I} \tag{5.30}$$

图 5.6 所示为电感两端的电压和电流随时间的变化，不难发现电感的电压超前电流 90°。

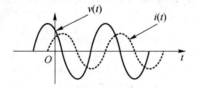

图 5.6　电感两端电压和电流随时间的变化

5.6　基尔霍夫定律的相量形式

根据第 2 章学习的基尔霍夫电流定律（KCL）和基尔霍夫电压定律（KVL），在本节将重写它们的相量形式。

5.6.1　KCL 的相量形式

根据第 2 章学习的 KCL，其相量形式可以表示为

$$\sum_{n=1}^{N}\dot{I}_n=0 \tag{5.31}$$

证明：

$$\begin{aligned}\sum_{n=1}^{N}i_n(t) &= \sum_{n=1}^{N}\mathrm{Re}[I_{mn}\mathrm{e}^{j\theta_n}\mathrm{e}^{j\omega t}] = \mathrm{Re}\Big[\mathrm{e}^{j\omega t}\cdot\sum_{n=1}^{N}I_{mn}\mathrm{e}^{j\theta_n}\Big] \\ &= \mathrm{Re}\Big[\mathrm{e}^{j\omega t}\cdot\sum_{n=1}^{N}\dot{I}_n\Big]=0\end{aligned} \tag{5.32}$$

由于 $e^{j\omega t} \neq 0$,因此 $\sum_{n=1}^{N} \dot{I}_n = 0$,即电路中任意节点所有流入电流(相量形式)的代数和为零。

5.6.2 KVL 的相量形式

根据第 2 章学习的 KVL,其相量形式可以表示为

$$\sum_{n=1}^{N} \dot{V}_n = 0 \tag{5.33}$$

证明:

$$\sum_{n=1}^{N} v_n(t) = \sum_{n=1}^{N} \text{Re}[V_{mn} e^{j\theta_n} e^{j\omega t}] = \text{Re}\left[\sum_{n=1}^{N} V_{mn} e^{j\theta_n} e^{j\omega t}\right] \tag{5.34}$$

$$= \text{Re}\left[e^{j\omega t} \cdot \sum_{n=1}^{N} \dot{V}_n\right] = 0$$

由于 $e^{j\omega t} \neq 0$,因此 $\sum_{n=1}^{N} \dot{V}_n = 0$,即在电路的任何环路上的所有电压(相量形式)的代数和为零。

5.7 电路简化方法的相量形式

5.7.1 串联阻抗和并联阻抗

串联阻抗可以通过将每个阻抗简单地相加而组合成一个阻抗。唯一的不同之处在于,将各个阻抗结合在一起会涉及复数的代数运算。

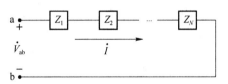

图 5.7 串联阻抗

如图 5.7 所示,a、b 端子之间的阻抗 $Z_1, Z_2, \cdots$ 和 Z_N 串联,因此它们有相同的电流 $\dot{I}$。根据 KVL 的相量形式,有

$$\dot{V}_{ab} = \dot{V}_1 + \dot{V}_2 + \cdots + \dot{V}_N = \sum_{n=1}^{N} \dot{V}_n = \sum_{n=1}^{N} (Z_n \dot{I}) = \dot{I} \sum_{n=1}^{N} Z_n \tag{5.35}$$

ab 端子之间的等效阻抗为

$$Z_{ab} = \frac{\dot{V}_{ab}}{\dot{I}_{ab}} = \sum_{n=1}^{N} Z_n \tag{5.36}$$

如图 5.8 所示,a、b 端子之间的阻抗 $Z_1, Z_2, \cdots, Z_N$ 并联,因此它们两端的电压 $\dot{V}_{ab}$ 是相同的。根据 KCL 的相量形式,有

$$i = \dot{I}_1 + \dot{I}_2 + \cdots + \dot{I}_N = \sum_{n=1}^{N} \dot{i}_n = \sum_{n=1}^{N}\left(\frac{\dot{V}_{ab}}{Z_n}\right) = \dot{V}_{ab}\sum_{n=1}^{N}\left(\frac{1}{Z_n}\right) \quad (5.37)$$

ab 端子之间的等效阻抗可表示为

$$\frac{1}{Z_{ab}} = \frac{\dot{I}}{\dot{V}_{ab}} = \sum_{n=1}^{N}\frac{1}{Z_n} \quad (5.38)$$

它可以重写为导纳的形式

$$Y_{ab} = \sum_{n=1}^{N} Y_n \quad (5.39)$$

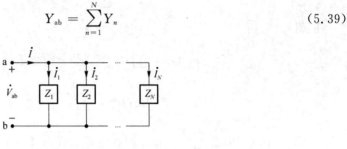

图 5.8　并联阻抗

5.7.2　电压源电流源的等效变换、戴维南定理和诺顿定理的相量形式

在第 3 章介绍的电压源电流源的等效变换、戴维南定理和诺顿定理也可以应用于复数域，写成相量表达形式。图 5.9 为戴维南定理与诺顿定理在复数应用电压源电流源的等效变换，可记作

$$\begin{cases} Z_N = Z_{TH} \\ \dot{V}_{TH} = Z_{TH}\dot{I}_N \end{cases} \quad (5.40)$$

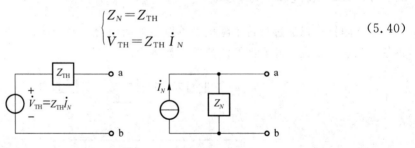

图 5.9　频域内的电压源电流源的等效变换

戴维南定理的相量表示是：一个含有独立电源和阻抗的线性网络的两端，其对外部而言可以用一个独立电压源串联一个阻抗来等效。其中，电压源的电压值 $\dot{V}_{TH}$ 等于 ab 两端开路时的电压，串联的阻抗值 Z_{TH} 等于网络内全部独立电源置零时网络 ab 端口的等效阻抗。

诺顿定理的相量表示是：一个含有独立电源和阻抗的线性网络的两端，其对外部而言可以用一个独立电流源并联一个阻抗来等效。其中，电流源的电流值 $\dot{I}_N$ 等于 ab 两端短路时的电流，并联的阻抗值 Z_N 等于网络内全部独立电源置零时网络 ab 端口的等效阻抗。

戴维南定理和诺顿定理的转化关系如式(5.40)所示。

5.8　电路基本分析方法的相量形式

5.8.1 节点电压法的相量形式

在第 2 章中,已经介绍了电路基础分析方法中的节点电压法,该方法的原理也可用于复数域电路的分析(相量形式)。下面,以一个例题为例,向大家解释节点电压法的相量形式。

问题 5.2 用节点电压法求出图 5.10 所示电路的电流 $\dot{I}_a$、$\dot{I}_b$ 和 $\dot{I}_c$。

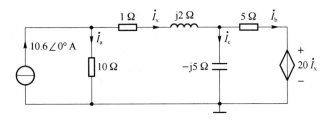

图 5.10 问题 5.2 中的电路

解:如图 5.11 标出节点电压。使用 KCL 和电路元件的 VCR,可以得到

$$\begin{cases} \dfrac{\dot{V}_1}{10} + \dfrac{\dot{V}_1 - \dot{V}_2}{1+\mathrm{j}2} - 10.6 = 0 \\ \dfrac{\dot{V}_2 - \dot{V}_1}{1+\mathrm{j}2} + \dfrac{\dot{V}_2}{-\mathrm{j}5} + \dfrac{\dot{V}_2 - 20\dot{I}_x}{5} = 0 \\ \dot{I}_x = \dfrac{\dot{V}_1 - \dot{V}_2}{1+\mathrm{j}2} \end{cases} \quad (5.41)$$

因此,节点电压为

$$\begin{cases} \dot{V}_1 = 68.4 - \mathrm{j}16.8 \text{ V} \\ \dot{V}_2 = 68 - \mathrm{j}26 \text{ V} \end{cases} \quad (5.42)$$

电流为

$$\begin{cases} \dot{I}_a = \dfrac{\dot{V}_1}{10} = 6.84 - \mathrm{j}1.68 \text{ A} \\ \dot{I}_x = \dfrac{\dot{V}_1 - \dot{V}_2}{1+\mathrm{j}2} = 3.76 + \mathrm{j}1.68 \text{ A} \\ \dot{I}_b = \dfrac{\dot{V}_2 - 20\dot{I}_x}{5} = -1.44 - \mathrm{j}11.92 \text{ A} \\ \dot{I}_c = \dfrac{\dot{V}_2}{-\mathrm{j}5} = 5.2 + \mathrm{j}13.6 \text{ A} \end{cases} \quad (5.43)$$

5.8.2 网孔电流法的相量形式

在第 2 章中,已经介绍了电路基础分析方法中的网孔电流法,该方法的原理也可用于复数域电路的分析(相量形式)。下面,以一个例题为例,向大家解释网孔电流法的相量形式。

问题 5.3 用网孔电流法求出图 5.12 所示电路中的电流 $i_1(t)$ 和 $i_2(t)$。

解:如图 5.13 标出网孔电流。把变量写成相量的形式:$\omega = 10^3$ rad/s,$\dot{V}_s = 10\angle 0°$,

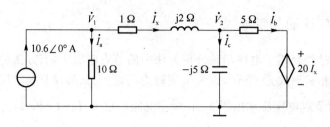

图 5.11 问题 5.2 中的电路加入节点电压

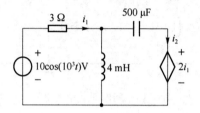

图 5.12 问题 5.3 中的电路

$Z_R = R = 3\ \Omega$, $Z_L = j\omega L = j10^3 \times 4 \times 10^{-3} = j4\ \Omega$ 以及 $Z_C = \dfrac{1}{j\omega C} = \dfrac{1}{j10^3 \times 500 \times 10^{-6}} = -j2\ \Omega$。

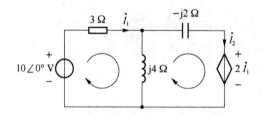

图 5.13 问题 5.3 中电路加入网孔电流

使用 KVL 和电路元件的 VCR,可以得到

$$\begin{cases} 3\dot{I}_1 + j4(\dot{I}_1 - \dot{I}_2) = 10\angle 0° \\ j4(\dot{I}_2 - \dot{I}_1) - j2\dot{I}_2 = -2\dot{I}_1 \end{cases} \tag{5.44}$$

因此,网孔电流为

$$\begin{cases} \dot{I}_1 = \dfrac{10}{7 - j4} = 1.24\angle 29.7°\ \text{A} \\ \dot{I}_2 = \dfrac{20 + j30}{13} = 2.77\angle 56.3°\ \text{A} \end{cases} \tag{5.45}$$

时域下的电流为

$$\begin{cases} i_1 = 1.24\cos(10^3 t + 29.7°)\ \text{A} \\ i_2 = 2.77\cos(10^3 t + 56.3°)\ \text{A} \end{cases} \tag{5.46}$$

5.9 功率计算

在第 1 章已经学习了功率的定义,并以直流电源为例介绍了功率的计算方法。由于交

流电路中变量的变化比直流电路要复杂,所以交流电路中的功率也不像直流电路那样简单。本节以正弦稳态电路为例,介绍与电路元件的功率相关的多个概念,具体包括瞬时功率、平均功率、无功功率、复功率和视在功率等。

在本章开始,已经介绍了正弦函数的均方根,即如果 $v(t)=V_\text{m}\cos(\omega t+\phi)$,则 $v(t)$ 的均方根为

$$V_\text{rms} = \sqrt{\frac{1}{T}\int_{t_0}^{t_0+T} V_\text{m}^2 \cos^2(\omega t+\phi)\text{d}t} = \frac{V_\text{m}}{\sqrt{2}} \tag{5.47}$$

其中,T 为正弦函数的周期。均方根是正弦信号的一个重要性质,且与下面的功率计算有关。

当电路中的电压或电源不再恒定,而是随时间变化时,功率也随着时间变化。为表征这一特点,功率被更加细致地定义为瞬时功率。

在电压电流关联参考方向下,元件或单口网络的**瞬时功率**定义为

$$\begin{aligned} p(t) &= v(t)i(t) = V_\text{m}\cos(\omega t+\theta_v-\theta_i) \cdot I_\text{m}\cos(\omega t) \\ &= \frac{V_\text{m}I_\text{m}}{2}\cos(\theta_v-\theta_i) + \frac{V_\text{m}I_\text{m}}{2}\cos(2\omega t+\theta_v-\theta_i) \\ &= \frac{V_\text{m}I_\text{m}}{2}\cos(\theta_v-\theta_i) + \frac{V_\text{m}I_\text{m}}{2}\cos(\theta_v-\theta_i)\cos(2\omega t) - \frac{V_\text{m}I_\text{m}}{2}\sin(\theta_v-\theta_i)\sin(2\omega t) \\ &= P + P\cos(2\omega t) - Q\sin(2\omega t) \end{aligned}$$
$$\tag{5.48}$$

其中,$P=\dfrac{V_\text{m}I_\text{m}}{2}\cos(\theta_v-\theta_i)$,$Q=\dfrac{V_\text{m}I_\text{m}}{2}\sin(\theta_v-\theta_i)$。

瞬时功率表示某个瞬间的功率,实际意义不大,也不便于测量,所以引入平均功率的概念。**平均功率**是瞬时功率在一段时间内平均值,记作 P

$$P = \frac{1}{T}\int_{t_0}^{t_0+T} p(t)\text{d}t = \frac{V_\text{m}I_\text{m}}{2}\cos(\theta_v-\theta_i) \tag{5.49}$$

式(5.49)的平均功率与式(5.48)中的 P 相同。平均功率也被称为**有功功率**,是电路元件或网络实际消耗或产生的功率。

重写公式(5.48),得

$$p(t) = P[1+\cos(2\omega t)] - Q\sin(2\omega t) \tag{5.50}$$

发现式(5.50)总是大于等于零,是瞬时功率中不可逆的部分,是消耗掉的功率;第二项是一个正弦函数,它的正负交替,是瞬时功率中的可逆部分,表示电路元件或网络与外加电源之间的能量交换。将这种能量交换的概念定义为**无功功率**,即无功功率

$$Q = \frac{V_\text{m}I_\text{m}}{2}\sin(\theta_v-\theta_i)$$

为了与有功功率区别,无功功率的单位定义为乏(Var)。

上面对功率进行了时域上的分析,当电压和电流用相量形式表示时,电压记作 $\dot{V}=V\angle\theta_v$,电流记作 $\dot{I}=I\angle\theta_i$,它们相乘记作

$$\dot{V}\dot{I} = V\angle\theta_v \cdot I\angle\theta_i = VI\angle(\theta_v+\theta_i) = VI\cos(\theta_v+\theta_i) + \text{j}VI\sin(\theta_v+\theta_i) \tag{5.51}$$

将式(5.51)与式(5.48)相比较,可看出它们的区别在于相位的相加和相减,在复数域这个问

题可通过共轭来解决。因此,定义**复功率**

$$\widetilde{S}=\dot{V}\dot{I}^*=V\angle\theta_v \cdot I\angle-\theta_i=VI\angle(\theta_v-\theta_i)=VI\cos(\theta_v-\theta_i)+jVI\sin(\theta_v-\theta_i)=P+jQ \tag{5.52}$$

由式(5.52)可知,复功率的实部为平均功率(或有功功率)P,虚部为无功功率Q。

复功率的模就是**视在功率**,记作

$$|\widetilde{S}|=VI=\frac{V_m I_m}{2}=\sqrt{P^2+Q^2} \tag{5.53}$$

其中,V 和 I 分别为电压和电流的有效值,视在功率的单位与复功率相同,为伏安(VA)。

不同功率及其单位的总结如表 5.2 所示。

表 5.2　不同功率及其单位的总结

功率类型	单　位
瞬时功率	伏安(VA)
复功率	伏安(VA)
视在功率	伏安(VA)
平均功率	瓦特(W)
无功功率	乏(Var)

5.10　最大功率传输定理的相量形式

在第 3 章已经学习了最大功率传输定理,该定理也可用于正弦稳态电路。

以图 5.14 所示的电路为例进行讨论,其中 $Z_s=R_s+jX_s$,$Z_L=R_L+jX_L$,则负载的复功率为

$$\widetilde{S}=\dot{V}\dot{I}^*=Z_L\dot{I}\dot{I}^*=|\dot{I}|^2(R_L+jX_L) \tag{5.54}$$

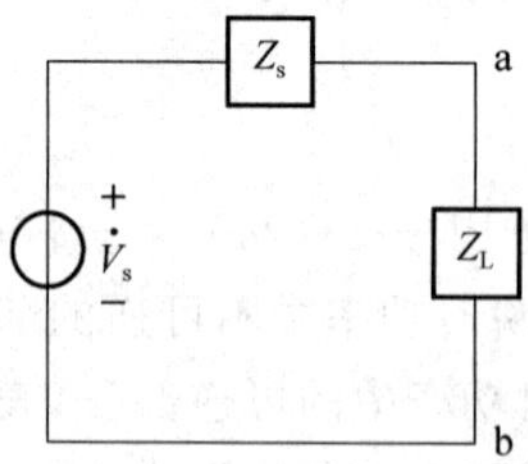

图 5.14　功率传输的相量形式

则负载的平均功率为

$$P_L=|\dot{I}|^2 R_L=\left|\frac{\dot{V}_s}{Z_L+Z_s}\right|^2 R_L=\left|\frac{\dot{V}_s}{(R_L+R_s)+j(X_L+X_s)}\right|^2 R_L=\frac{\dot{V}_s^2 R_L}{(R_L+R_s)^2+(X_L+X_s)^2} \tag{5.55}$$

为最大化负载 Z_L 获得的功率,令

$$\begin{cases} \dfrac{\partial P_L}{\partial X_L}=0 \\ \dfrac{\partial P_L}{\partial R_L}=0 \end{cases} \tag{5.56}$$

可得 $\begin{cases} X_L=-X_s \\ R_L=R_s \end{cases}$,即 $Z_L=Z_s^*$。

由此可知,当负载阻抗为电源内阻阻抗的共轭时 $Z_L=Z_s^*$,负载可获得最大功率,

$$P_{L,\max}=\dfrac{\dot V_s^2}{4R_s} \tag{5.57}$$

这也被称为共轭匹配。

5.11 习　　题

习题 5.1　一个 $5\ \mu F$ 电容两端的电压为 $30\cos(4\,000t+25°)\ V$。计算:(1) 电容的电抗;(2) 电容的阻抗;(3) 相量电流 $\dot I$;(4) 电流 $i(t)$ 的稳态表达式。

习题 5.2　假设 $v(t)=20\cos(\omega t+15°)$ 而 $i(t)=4\sin(\omega t-15°)$,计算网络两端的平均功率和无功功率,并说明该网络是吸收功率还是释放功率。

习题 5.3　已知正弦电压 $v(t)=100\sin(628t-30°)V$,试求电压的最大振幅、有效值、角频率、周期、相位角(弧度)。

习题 5.4　假设正弦电压 $v(t)=180\cos(20\pi t-60°)V$,试求:

(1) 电压的最大振幅是多少?

(2) 频率(Hz)是多少?

(3) 频率(弧度/秒)是多少?

(4) 相位角(弧度)是多少?

(5) 相位角(度)是多少?

(6) 周期是多少秒?

习题 5.5　若正弦电压 $v_1(t)=60\sin(\omega t-30°)\ V$,$v_2(t)=10\cos\omega t\ V$,试判断二者的相位关系。

习题 5.6　图 5.15 所示正弦交流电路中,已知 $\dot V=12\angle0°\ V$,$\dot I=5\angle-36.9°\ A$,$R=3\ \Omega$,求 $\dot I_L$ 及 ωL。

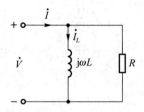

图 5.15 习题 5.6 中的电路

习题 5.7 图 5.16 所示正弦交流电路中,已知电流 $\dot{I}$ 的有效值为 10 A,电流 $\dot{I}_2$ 的有效值为 6 A,求电流 $\dot{I}_1$。

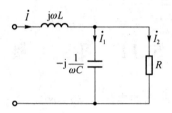

图 5.16 习题 5.7 中的电路

习题 5.8 图 5.17 所示正弦交流电路中,已知电流 $\dot{I}_1$ 的有效值为 4 A,电流 $\dot{I}_2$ 的有效值为 3 A,求电流 $\dot{I}$。

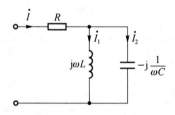

图 5.17 习题 5.8 中的电路

习题 5.9 图 5.18 所示正弦交流电路中,已知电流有效值分别为 $I=5$ A, $I_R=5$ A, $I_L=3$ A,求 $\dot{I}_C$。若 $I=5$ A, $I_R=4$ A, $I_L=3$ A,此时 $\dot{I}_C$ 又为多少?

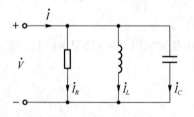

图 5.18 习题 5.9 中的电路

习题 5.10 图 5.19 所示为某正弦交流电路的一部分,已知 $\dot{I}_R=2\angle-\dfrac{\pi}{3}$ A,求 $\dot{I}_L$。

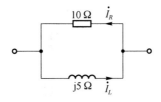

图 5.19 习题 5.10 中的电路

习题 5.11 求图 5.20 所示正弦交流电路中的电流 $\dot{I}$。

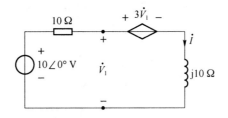

图 5.20 习题 5.11 中的电路

习题 5.12 图 5.21 所示电路中的相量电流 $\dot{I}_b$ 为 $5\angle 45°$。

(1) 计算 $\dot{I}_a$、$\dot{I}_c$ 和 $\dot{V}_g$；

(2) 假设 $\omega=800$ rad/s，写出 $i_a(t)$、$i_c(t)$ 和 $v_g(t)$ 的表达式。

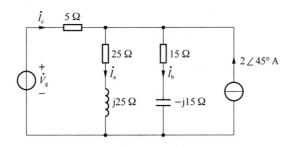

图 5.21 习题 5.12 中的电路

习题 5.13 求图 5.22 所示单口网络的导纳 Y。

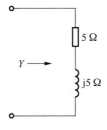

图 5.22 习题 5.13 中的电路

习题 5.14 图 5.23 所示 RL 串联电路中，已知 $R=1\ \Omega$，$L=1$ H，$\omega=1$ rad/s，求该电路的阻抗 Z。

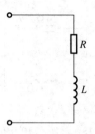

图 5.23 习题 5.14 中的电路

习题 5.15 求图 5.24 所示电路分别在 $\omega=0$ 和 $\omega=\infty$ 时的阻抗 Z_{ab}。

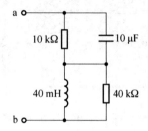

图 5.24 习题 5.15 中的电路

习题 5.16 图 5.25 电路中的正弦电压源产生的电压等于 $247.49\cos(1\,000t+45°)$ V。
(1) 求出 a、b 端子间的戴维南等效电压；
(2) 求出 a、b 端子间的戴维南等效阻抗；
(3) 画出 a、b 端子间的戴维南等效电路。

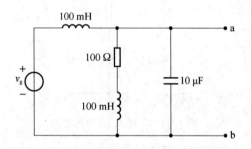

图 5.25 习题 5.16 中的电路

习题 5.17 图 5.26 所示正弦交流电路中，已知 $v(t)=30\sqrt{2}\sin(\omega t-30°)$ V, $\omega=10^3$ rad/s, 求 $i_1(t)$、$i_2(t)$、$i_3(t)$ 和 $i(t)$。

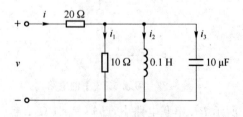

图 5.26 习题 5.17 中的电路

习题 5.18　如图 5.27 所示，假设 i_g 等于 $30\cos100t$ mA，求出电路中负载的平均功率、无功功率和视在功率。

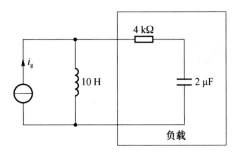

图 5.27　习题 5.18 中的电路

题 5.19　求图 5.28 所示正弦交流电路中电源供出的平均功率。

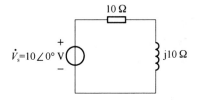

图 5.28　习题 5.19 中的电路

题 5.20　图 5.29 所示正弦交流电路中，已知 $R=\omega L=\dfrac{1}{\omega C}=100\ \Omega$，$\dot{I}_R=2\angle0°$ A。求 $\dot{V}_s$ 和电源供出的有功功率。

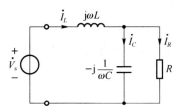

图 5.29　习题 5.20 中的电路

第6章 实　验

针对本书前述知识点，本章将通过软件或硬件实验来验证在前面几章所学到的电路分析方法和电路定理。

本章的目标：
(1) 学习使用基本的电路测量仪器；
(2) 学习使用软件进行电路仿真；
(3) 验证电路分析方法和电路定理。

6.1　基本电路测量仪器介绍

6.1.1　万用表

万用表(伏特-欧姆-毫安表)是一种电子测量仪器(如图 6.1 所示)，它在一个设备里集成了多种测量功能。一个典型的万用表可以测量电压、电流和电阻。万用表可以是一种手持式设备，用于基本的故障查找和现场检测，也可以是一种台式设备，以测得非常高的精度。

(a) 模拟万用表　　(b) 数字万用表

图 6.1　万用表

万用表的使用方法如下:

(1) 在使用万用表之前,应先进行"机械调零",即在没有被测电量时,万用表指针指在零电压或零电流的位置上;

(2) 在使用万用表过程中,不能用手去接触表笔的金属部分,这样一方面可以保证测量的准确,另一方面也可以保证人身安全;

(3) 在测量某一电量时,不能在测量的同时换挡,尤其是在测量高电压或大电流时,更应注意,否则,会使万用表损坏,如需换挡,应先断开表笔,换挡后再去测量;

(4) 万用表在使用时,必须水平放置,以免造成误差,同时,还要注意到避免外界磁场对万用表的影响。

6.1.2 直流电源

电源是一种为电子负载提供电能的电子设备。电源的主要功能是将电流从源端转换为正确的电压、电流和频率,从而为负载供电。因此,电源有时被称为电力转换器。有些电源是独立的设备,而有些则内置在它们所供电的负载设备中。电源的其他功能还包括限制流经负载的电流不超过某个安全值,在发生电气故障时的停止电能供应,功率调节(防止电子噪音或到达负载的电压激增),功率因素校正,以及为了能够在发生临时断电时可以继续供电的电能储存(不间断电源)。

直流电源是一种为负载提供恒定直流电压的电源。根据其结构设计,直流电可以由直流电源供电,也可以由交流电源供电,如电源干线。

6.1.3 电阻和滑动变阻器

电阻器是一种无源双端电子元件,它将电子学上的电阻作为电路元件来实现。在电子电路中,电阻用于减小电流、调整信号电平、分压、偏置有源元件和终止传输线等用途。固定电阻的阻值只随温度、时间或工作电压有轻微的变化。可变电阻可用于调整电路元件(如音量控制或调光器),或作为热、光、湿度、力或化学活性的传感装置。

电阻的功能是由它的阻值决定的:普通的商用电阻的阻值范围超过 9 个数量级。电阻的标称值在其制造公差范围内,并在元件上注明。

色环电阻是电路中常见的分立元件,采用色环来表示颜色和误差,这类电阻的基本单位是:欧(Ω)、千欧($k\Omega$)、兆欧($M\Omega$)。规定的颜色为黑、棕、红、橙、黄、绿、蓝、紫、灰、白、金、银、无色带。表 6.1 为各颜色代表的数值。

表 6.1 电阻色环颜色代表的数值

	银	金	黑	棕	红	橙	黄	绿	蓝	紫	灰	白	无
有效数字	—	—	0	1	2	3	4	5	6	7	8	9	—
数量级	10^{-2}	10^{-1}	10^{0}	10^{1}	10^{2}	10^{3}	10^{4}	10^{5}	10^{6}	10^{7}	10^{8}	10^{9}	—
允许偏差(%)	±10	±5	—	±1	±2	—	—	±0.5	±0.25	±0.1	±0.05	—	±20

识别方法：

（1）三色环电阻：第一色环是十位数，第二色环是个位数，第三色环代表倍率。用前三个色环来代表其阻值。

（2）四色环电阻：第一、二环分别代表两位有效数的阻值；第三环代表倍率；第四环代表误差。

（3）五色环电阻：第一、二、三环分别代表三位有效数的阻值；第四环代表倍率；第五环代表误差。如果第五条色环为黑色，一般用来表示为绕线电阻器；如果第五条色环如为白色，一般用来表示为保险丝电阻器。

（4）六色环电阻：六色环电阻前五色环与五色环电阻表示方法一样，第六色环表示该电阻的温度系数。

变阻器是一种在断开电路的情况下改变电阻阻值的装置。人们最熟悉的变阻器的形式可能是一个用来改变光的强度的调光器或滑块。在这里，变阻器被用来设置舒适的照明水平。从而使人们在不需要更换灯的前提下，改变光照的水平。变阻器也被应用于许多电气应用和各种工业场景。

这种元件是基于一个现象，即流过电路的电流会随着它所遇到的总电阻的大小变化而变化。低电阻意味着高电流，高电阻则意味着低电流。电路的这种特性可以被应用于电路性能的改变，以满足特定的需要。

6.1.4 电容和电感

电容是一种能在电场中储存电能的无源双端电子元件。考虑到实验安全，需注意电解电容在以下情况下容易爆炸：一是极性接反，二是电压超高。

电感器是一种无源双端电子元件，当电流流过它时，它将能量储存在磁场中。电感器通常由绕在铁芯周围的线圈中的绝缘导线组成。

6.1.5 示波器

示波器（以前称为记录仪）是一种通过图形的方式（通常是一个描绘一个或多个信号的时间函数的二维图）显示不同的信号电压的电子测量仪器。其他信号（如声音或振动）可以转换成电压并显示出来。示波器在校准刻度上显示一个电信号随时间的变化，其中 Y 轴和 X 轴分别表示电压和时间。然后便可以分析波形的振幅、频率、上升时间、时间间隔、失真等特性。现代数字仪器可以直接计算和显示这些特性。

可以对示波器进行调节，从而使重复的信号在屏幕上形成一个连续的图形，进而观察。存储示波器可以捕获单个的结果并持续地显示它，从而使用户可以更好地观察它。否则，它的显示时间太短，用户无法观察。

6.1.6 数字信号发生器

数字信号发生器是在数字域中产生重复或非重复电子信号的电子设备。不同类型的信号发生器，具有不同的用途和应用，并且成本也不同。数字信号发生器包括函数发生器、射频和微波信号发生器、螺距发生器、任意波形发生器、数字模式发生器和频率发生器。一般来说,没有一种设备能够适应所有可能的应用。常用的一种数字信号发生器为函数发生器。它是一种能产生简单重复波形的装置。

这些设备包含一个电子振荡器和一个能够产生重复波形的电路。现代设备可以使用数字信号处理来合成波形,然后使用数模转换器(DAC)来产生模拟输出。最常见的波形是正弦波,此外,阶跃(脉冲)、方波、三角波以及任意波形也都通常是能实现的。

6.2 电路仿真软件介绍

接下来介绍一个电路模拟器集合,它同时可以用于电路绘图、电路设计以及电路分析。下面,总结了几个常用的电路仿真软件。

MultiSim 是一个由国家仪器开发的电路仿真软件。众所周知,学生版本的访问权限往往是受约束的。但它对于电子初学者来说,仍然是一个很棒的模拟工具。MultiSim 能够捕获电路,创建布局,分析电路和仿真。它的重点功能包括在提交实验任务之前,使用 3D 环境模拟面包板,创建印刷电路板(PCB)等。使用 Multisim 电路仿真器可以实现面包板仿真。

NgSpice 是 Sourceforge 开发的广泛使用的免费开源电路模拟器。Ngspice 是 gEDA 项目的一部分,该项目每天都根据其用户提供的建议进行的改进以及 bug 的修复。由于这是一个合作项目,因此用户可以对电路模拟器提出改进建议,并成为开发团队的一部分。

除此之外,还有一些软件大家可以自行学习和安装。

6.3 实验实例

6.3.1 电阻的直流特性

1. 实验目的
- 学习使用万用表和直流电压源;
- 搭建、测量并分析简单的无源电路元件。

2. 实验设备

本节实验将研究如何连接简单的电路,如何根据电路原理图构建电路,以及如何使用万用表对每个元件的电阻进行简单的测量。实验设备包括直流电源、万用表、电阻器、滑动变阻器和导线。

3. 实验方法、原理或规律
- 欧姆定律指出,在同一电路中,通过电阻的电流与电阻两端的电压成正比,如下所示。

$$I = \frac{V}{R} \tag{6.1}$$

其中,I 是通过导体的电流,单位是安培(A);V 是导体两端测得的电压,单位是伏特(V);R 是导体的电阻,单位是欧姆(Ω)。此外,欧姆定律指出,式(6.1)中的导体电阻 R

是常数,与通过它的电流无关。
- 逐点检测法是使用一个参数作为参考,通过改变次标准的数值,从而测量并检查相关变量。
- 电阻值

Colour	code		
Brown	1		
Red	2		
Orange	3		
Yellow	4		
Green	5	Colour	Tolerance
Blue	6	Gold	5%
Violet	7	Silver	10%
Grey	8	Colouruless	20%
White	9		
Black	0		

4. 注意事项

更换元件之前,请先关闭直流电源。

5. 通过 I-V 数据测电阻

测量 $R=200\ \Omega$ 电阻的伏安特性。

(1) 根据图 6.2 所示电路图连接电路。

(2) 改变直流电压电源的值,用万用表测量相应的电流 I、电阻两端的电压 V,并将结果记录在表 6.2 中,并计算出 R_{th} 的理论值。

(3) 用万用表测量 R 值,并与理论 R_{th} 值进行比较。

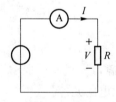

图 6.2 电阻阻值测量电路图

表 6.2 电阻阻值测量结果

实验标号	V/V	I/A	R_{th}/Ω
1			
2			
3			

$R=$ _____

6. 探究电阻的串/并联特性

(1) 串联电路

分别测量 $R_1=200\ \Omega$ 电阻和滑动变阻 R_2 的各自的以及其总和的伏安特性。图 6.3 中的电路示意图显示了几个电阻之间的连接。

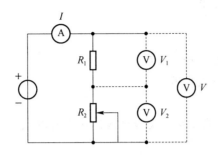

图 6.3　电阻串联电路图

改变滑动变阻器滑块的位置,用万用表测量对应的阻值 R_1 和 R_2,以及电压 V_1 和 V_2,将总电压 V 和电路电流 I 记录在表 6.3 中。

表 6.3　电阻串联特性测量结果

实验标号	V_1/V	V_2/V	V/V	R_1/Ω	R_2/Ω	I/A
1						
2						
3						

(2) 并联电路

分别测量电阻 $R_0=200\ \Omega$ 和 $R_1=200\ \Omega$ 以及滑动变阻器 R_2 的各自的以及其总和的伏安特性。图 6.4 中的电路示意图描绘了几个电阻之间的连接。

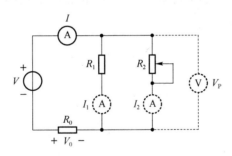

图 6.4　电阻并联电路图

改变滑动变阻器滑块的位置,用万用表测量相应的阻值 R_2、电压 V_0 和 V_p,电路电流 I_1、I_2 和 I,并记录在表 6.4 中。

表 6.4　电阻并联特性测量结果

实验标号	V_0/V	V_p/V	V/V	R_2/Ω	I/A	I_1/A	I_2/A
1							
2							
3							

6.3.2 验证基尔霍夫定律

1. 实验目的
- 使用实验数据验证基尔霍夫定律；
- 深化对电路参考方向的理解。

2. 实验设备

这个实验将验证基尔霍夫定律。实验设备包括直流电源、万用表、电阻、滑动变阻器和导线。

3. 实验方法、原理或规律

基尔霍夫定律是一种最基本的电路理论。基尔霍夫定律有两个：一个是电流定律，另一个是电压定律。

(1) 基尔霍夫电流定律(KCL)：在任一时刻，电路的任一节点，进入该节点的电流的代数和为零，即

$$\sum_{n=1}^{N} i_n(t) = 0 \tag{6.2}$$

$$i_A + i_B + (-i_C) + (-i_D) = 0$$

请注意，电流的参考方向对于 KCL 是至关重要。尽管电流的参考方向可以任意设定，但是在之后的分析中需要遵从设定的参考方向。

(2) 基尔霍夫电压定律(KVL)：在任一时刻，电路的任一回路中，各支路电压代数和为零，即

$$\sum_{n=1}^{N} v_n(t) = 0 \tag{6.3}$$

$$-v_1 + v_2 - v_3 = 0$$

请注意，在 KVL 中参考方向和参考极性都需考虑。

4. 防范措施

(1) 注意电流表和电压表的极性；
(2) 合理选择量程，不要超过量程使用仪表；
(3) 稳压电源的输出要由小至大进行调节。

5. 基尔霍夫电压定律

根据图 6.5 测量每个点之间的电位差和两个点的顺序，并将测量的数据填写在表 6.5 中。

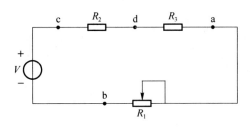

图 6.5　电阻串联电路图

表 6.5　基尔霍电压定律测量结果

电　压		V_{cd}	R_2	V_{da}	R_3	V_{ca}	V_{ab}	R_1	V_{cb}	V
$R_1=$_____	测量值									
$V=$_____	计算值		/		/			/		
$R_1=$_____	测量值									
$V=$_____	计算值		/		/			/		

结果分析：

（1）测量值 V_{cd}、V_{da}、V_{ab} 和电压 V 之间的关系是什么？

（2）试着分析误差，$\left(\text{误差}=\dfrac{\text{测量值}-\text{计算值}}{\text{计算值}}\right)$

（3）结论：

6. 基尔霍夫电流定律

在图 6.6 中，V 代表稳定的独立电压源，R_1、R_2、R_3、R_4、R_5 代表定值电阻。

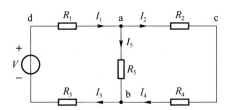

图 6.6　电阻并联电路图

表 6.6　基尔霍夫电流定律测量结果

电流名称	I_1	I_2	I_3	I_4	I_5
测量值					
计算值					

结果分析：

(1) 测量值 I_1、I_2 和 I_3 之间的关系是什么呢？

(2) 能在图 6.6 中找到另一组电流来验证 KCL 吗？证明它。

(3) 试着去分析误差。

(4) 结论：

6.3.3 RC 电路和 RL 电路的测量

1. 实验目的

- 学习使用数字信号发生器和示波器；
- 理解 RC 和 RL 电路和积分电路的概念。

2. 实验设备

本实验将介绍如何测量 RC 和 RL 电路。实验设备包括直流电源、数字信号发生器、示波器、电阻器、电容、电感器、LED 和导线。

3. 实验方法、原理或规律

(1) RC 电路（电阻-电容电路）

RC 电路是指由电阻元件和电容元件组成的一阶电路。最简单的 RC 电路就是一个电容和一个电阻串联（如图 6.7 所示）。当电路只由一个充电电容和一个电阻组成时，电容将通过电阻释放其储存的能量。

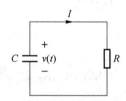

图 6.7 一个简单 RC 电路图

电容两端的电压与时间有关，其值可以用基尔霍夫电流定律求解，其中流过充电电容的电流必须等于通过电阻的电流。这就得到了线性微分方程：

$$C\frac{\mathrm{d}v(t)}{\mathrm{d}t}+\frac{v(t)}{R}=0 \tag{6.4}$$

其中，C 为电容的容值。

求解这个关于 $v(t)$ 的方程，得到指数衰减公式：

$$v(t)=V_0\mathrm{e}^{-\frac{t}{RC}} \tag{6.5}$$

其中，V_0 为 $t=0$ 时电容两端的电压。

(2) RL 电路（全称电阻-电感电路）

RL 电路是指由电阻元件和电感元件组成的一阶电路。它由一个电阻器、一个电感元件串联或并联组成。最简单的 RL 电路就是一个电感和一个电阻串联（如图 6.8 所示）。

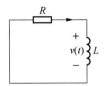

图 6.8　一个简单 RL 电路图

电感两端的电压也可以用基尔霍夫电流定律求出来，

$$\frac{1}{L}\frac{\mathrm{d}v(t)}{\mathrm{d}t}+\frac{v(t)}{R}=0 \tag{6.6}$$

然后

$$v(t)=V_0 \mathrm{e}^{-\frac{tR}{L}} \tag{6.7}$$

其中，V_0 为 $t=0$ 时电感两端的电压。

(3) 积分电路

积分电路是使输出信号与输入信号的时间积分值成比例的电路。最简单的积分电路由一个电阻和一个电容构成，如图 6.9 所示。其中，电流 I 的计算公式为

$$\dot{I}=\frac{\dot{V}_{\mathrm{in}}}{R+\dfrac{1}{\mathrm{j}\omega C}} \tag{6.8}$$

电容两端的电压 V_C 为

$$V_C=\frac{1}{C}\int_0^t I \mathrm{d}t \tag{6.9}$$

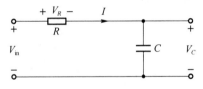

图 6.9　一个简单 RC 积分电路图

考虑高频电容的输出，

$$\omega \gg \frac{1}{RC} \tag{6.10}$$

因此

$$I \approx \frac{V_{\mathrm{in}}}{R} \tag{6.11}$$

则

$$V_C \approx \frac{1}{RC}\int_0^t V_{\mathrm{in}} \mathrm{d}t \tag{6.12}$$

4. 分别观察 *RC* 和 *RL* 电路中的 LED 灯

(1) 观察 *RC* 电路中的 LED 灯

如图 6.10 所示连接电路，其中 $R=200\ \Omega$ 且 $V=10\ \mathrm{V}$，然后把直流电压源移除，观察 LED 灯并记录现象。

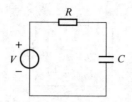

图 6.10　一个简单 RC 电路图

（2）观察 RL 电路中的 LED 灯

如图 6.11 所示连接电路，其中 $R=200\ \Omega$ 且 $V=10\ \text{V}$，然后把直流电压源移除，观察 LED 灯并记录现象。

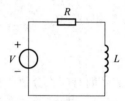

图 6.11　一个简单 RL 电路图

5. RC 积分电路的输入与输出

如图 6.12 所示连接电路，其中 V_{in} 为正弦波 $V_{\text{in}}=V_{\text{m}}\cos(\omega t)(V_{\text{m}}=5\ \text{V})$。随着时间的变化，观察输出 V_C 并将 V_{in} 和 V_C 随时间变化的图像画出来。

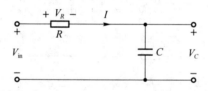

图 6.12　一个简单 RC 积分电路图

画出输出波形的图像。

6.3.4　验证戴维南定理的实验

1. 实验目标

- 掌握戴维南等效电路；
- 验证戴维南定理，并加深对该定理的理解；
- 学习如何测量网络的等效参数。

2. 实验设备

在本次实验中，将学习如何测量有源单端口网络的等效参数，以及如何验证戴维南定理。实验设备包括直流电源、万用表、电阻器和电线。

3. 实验方法、原理或规律

（1）戴维南电路和电路参数

戴维南定理指出：对于任何线性有源单端口网络都可以使用一个独立电压源串联一个

电阻来等效。该电压源的电压值 V_s 等于该有源网络 ab 两端开路时的电压 V_{oc}，该串联电阻的阻值 R_0 等于网络内全部独立电源置零时该网络 ab 两端的等效电阻。

如图 6.13 所示为有源单端口网络 ab 两端的戴维南等效电路。

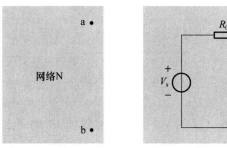

图 6.13 戴维南定理图示

(2) 有源单端口网络等效电阻的测量方法

① 用伏安法测量 R_0

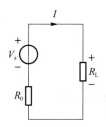

图 6.14 电阻串联电路图

如图 6.14 所示，使用万用表测量电阻 R_L 两端的电压和通过它的电流，并得到该网络的外部特性曲线如图 6.15 所示。根据外部特性曲线计算斜率 $\text{tg}\,\phi$，则内阻为

$$R_0 = \text{tg}\,\phi = \frac{\Delta V}{\Delta I} = \frac{V_{oc}}{I_{sc}} \tag{6.13}$$

若网络的内阻值很低，则不宜测其短路电流。

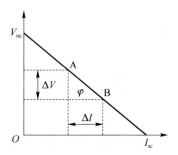

图 6.15 戴维南等效电路的伏安特性曲线

② 半电压法

如图 6.16 所示，当负载电压为被测网络开路电压 V_{oc} 的一半时，断开电路中的电阻 R_L，并用万用表测量其阻值。此时，R_L 的阻值即为等效电路的阻值 R_0。

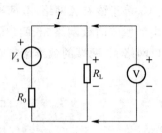

图 6.16 电阻串联电路

(3) 负载最大功率的条件

图 6.14 可以视作是一个由电源向负载供能的模型,其中 R_0 可以看作是电源电阻和导线电阻之和,R_L 为网络两端的负载。负载 R_L 上的功率消耗 P 可以由如下的公式表示:

$$P = I^2 R_L = \left(\frac{V_s}{R_0 + R_L}\right)^2 R_L \tag{6.14}$$

其中,$R_L = 0$ 或 $R_L = \infty$,电源传输给负载的功率为零。改变上式中 R_L 电阻的阻值,可以得到不同的功率值 P,其中必定会有一个 R_L 值,使得负载功率达到最大。

根据最大功率传输定理可知,当 $R_L = R_0$ 时,负载从电源获得的功率最大为

$$P_{\max} = \left(\frac{V_s}{R_0 + R_L}\right)^2 R_L = \left(\frac{V_s}{2R_L}\right)^2 R_L = \frac{V_s^2}{4R_L} \tag{6.15}$$

此时,电路处于"匹配"状态。

4. 用伏安法测量戴维南等效电路

(1) 实验电路如图 6.17 所示;

(2) 调整直流稳压电源,使输出电压 $V_s = 12$ V,且保持不变;

(3) 移除负载 R_L,测量网络两端的开路电压 V_{oc},并做好记录;

(4) 移除负载 R_L,并测量该支路的短路电流 I_{sc},并记录数据;

(5) 连接负载 R_L,根据表中要求改变负载 R_L,并测量不同电阻 R_L 对应的 V_{ab} 和 I,并记录数据(表 6.7);

(6) 利用表中数据,绘制网络 N 的伏安特性曲线 $V_{ab} = f(I)$,用伏安法计算 R_0;

(7) 根据开路电压 V_{oc} 和负载 R_0,得到单端口网络 N 的戴维南等效电路。

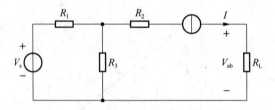

图 6.17 用伏安法测量戴维南等效电路

表 6.7 用伏安法测量戴维南等效电路测量结果

R_L/Ω	100	200	270	470	600	1 000	开路
V_{ab}/V							
I/mA							

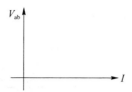

根据画出来的图计算 $R_0=$ _____ Ω, $V_{oc}=$ _____ V。

5. 用半电压法测量诺顿等效电路

(1) 实验电路如图 6.18 所示;
(2) 测量上述电路的短路电流 I_{sc};
(3) 测量上述电路开路电压 V_{oc};
(4) 根据半电压法,调整滑动变阻器,当电压为 V_s 的一半时,断开电路中的电阻 R_L,并用万用表测量阻值。R_L 的阻值为等效电路的阻值 R_0;
(5) 改变滑动变阻器的阻值,并测量不同的阻值下的电压 V_{ab} 和 I'。记录数据(表 6.8)并绘制伏安特性曲线;
(6) 根据短路电流 I_{sc} 和 R_0,得到有源单端口网络 N 的诺顿等效电路。

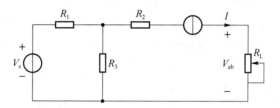

图 6.18 用半电压法测量诺顿等效电路

表 6.8 用半电压法测量诺顿等效电路测量结果

R_L/Ω	100	200	270	470	600	1 000	开路
V_{ab}/V							
I/mA							

$R_0=$ _____ Ω, $I=$ _____ mA。

Forward

This book is mainly for the undergraduates who are new to circuit knowledge. The goal of this book is to explain the basic concepts and analysis methods in circuit courses. The bilingual version is suitable for international students and domestic students who intend to study abroad in the future. Through this book, students can systematically learn basic electronic components, circuit analysis principles and methods, especially including some basic circuit theorems and circuit techniques. In addition, this book also provides four basic experiments related to some basic knowledge introduced in this book. We hope that students develop their scientific thinking ability, equip with the preliminary engineering evaluation ability, and lay a good foundation for the follow-up circuit courses (such as electronic circuit basics and digital circuit design), with the help of this book.

The book has six chapters, respectively introducing the basic circuit concepts, basic circuit analysis methods, basic circuit theorems, and the effects of different components and different types of signals on circuit performance. The keyword of this book is basic circuit knowledge. Through summarization and horizontal and vertical comparison, the book helps readers understand the abstract circuit knowledge better. Following summarizes the goals of all chapters, which helps readers focus on the key knowledge.

The goal of Chapter 1: Basic circuit concept

(1) understand the definition and the relationship between charge, voltage, current, power and energy are provided;

(2) understand the passive sign convention and skill in power calculation;

(3) understand resistor, capacitor, inductor, independent source and dependent source;

(4) skill in the voltage-current relationship (VCR) of resistor, capacitor and inductor.

The goal of Chapter 2: Basic circuit analysis method

(1) understand the definition of nodes, paths, loops, branches and meshes;

(2) skill in the equivalence of series and parallel combinations;

(3) understand and skill in Kirchhoff's Current Law (KCL) and Kirchhoff's Voltage Law (KVL);

(4) understand 2b-method;

(5) skill in node-voltage method;

(6) skill in mesh-current method.

The goal of Chapter 3: Basic circuit theorem

(1) understand and skill in the superposition theorem;

(2) understand and skill in source transformation;

(3) understand and sill in Thevenin theorem and Norton theorem;

(4) understand and sill in maximum power transfer.

The goal of Chapter 4: First-order circuit

(1) understand the characteristics of inductor and capacitor, skill in their VCR and the equivalence of series and parallel combinations;

(2) understand switching rules;

(3) understand the definition of first-order circuit, zero-input and zero-state response of RL and RC circuit;

(4) skill in calculating zero-input response of RL and RC circuit;

(5) skill in calculating zero-state response of RL and RC circuit;

(6) understand the instantaneous and the steady response, and skill in calculating the complete response and three-factor method.

The goal of Chapter 5: Sinusoidal steady state analysis

(1) understand the definition of sinusoidal signal, sinusoidal source and sinusoidal response;

(2) skill in the phase representation of sinusoidal variable;

(3) understand of the definition of impedance and admittance;

(4) understand and skill in the analysis methods and theorems in phase representation;

(5) understand the complex power and its variants, and skill in their calculations.

The goal of Chapter 6: verify the circuit analysis methods and circuit laws studied in the previous chapters through software or hardware experiments

(1) learn to use basic circuit measuring instruments;

(2) learn to use software for circuit simulation;

(3) verify the circuit analysis methods and circuit laws.

Please think about the purpose of the circuit analysis. First, we talk about what is circuit analysis? When we study the circuit theory, we commonly divided it into two parts, i.e., circuit analysis and circuit design. As Fig. 0.1 shows, the difference between circuit analysis and circuit design is obvious. Circuit analysis refers to calculating responses of a circuit with known input signals, circuit topology and parameters, while circuit design refers to the design of circuit topology and parameters with known input signals and output signals. The goal of circuit analysis is to analyze circuit variables quantitatively, meanwhile it also provides suggestions for circuit design improvements. In this book, we mainly focus on the introduction of the knowledge involved in circuit analysis and the explanation of analysis methods.

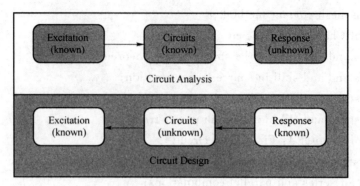

Fig. 0.1 The categories of circuit theory

Circuit analysis is based on mathematical techniques and is used to predict the circuit model and parameters, which plays a very important role in the design process. Specifically, this technique uses mathematical models to establish equations for a given circuit, and then solves the equations to determine the current and voltage required in the circuit. When establishing the equation, the circuit equation is mainly written based on topological constraints and component constraints. Topological constraints represent the connectivity of a circuit, and component constraints describe the VCR of circuit elements.

Chapter 1

Circuit Elements and Circuit Variables

Circuit analysis refers to the process of quantitative analysis of circuit variables. In this chapter, we first introduce the basic knowledge of circuit theory, circuit components and circuit variables.

The goal of Chapter 1: Basic circuit concept

(1) understand the definition and the relationship between charge, current, voltage, power and energy;

(2) understand the passive sign convention and skill in the power calculation;

(3) understand resistor, capacitor, inductor, independent source and dependent source;

(4) skill in the voltage-current relationship (VCR) of resistor, capacitor and inductor.

1.1 Common Circuit Variables

This section introduces some common and basic circuit variables, including charge, voltage, current, power, and energy, as well as their calculation methods and conversion relationships.

1.1.1 Electric charge

Electric charge is the basis for describing all electrical phenomena. Electrical effect is from electric charge moving and separation. Electric charge is bipolar. According to the directionality of the electric force, electric charge can be divided into positive charge and negative charge. In high school physics, we have learned the difference of positive charge and negative charge, i.e., positive charge is produced by rubbing glass with silk and negative charge is produced by rubbing amber with fur.

In physics, charge quantity is the amount of electric charge, which can also be called electric charge for short. In the International System of Units, charge quantity is denoted by Q, and the unit is Coulomb (C). The elementary charge (the charge of the proton) is exactly $1.602\,176\,634 \times 10^{-19}$ coulombs. Thus the coulomb is exactly the charge of

$1/(1.602\,176\,634\times10^{-19})$ protons, The same number of electrons has the same magnitude but opposite sign of charge, that is, a charge of 1 C.

1.1.2 Voltage

Voltage, also known as potential difference, is a physical quantity that measures the energy difference in electric potential between two points. The difference in electric potential between two points (i.e., voltage) in a static electric field is defined as the work needed per unit of charge to move a test charge between the two points. The direction of the voltage is defined as the direction from high electric potential to low electric potential. The unit of voltage is volts (V for short). Commonly used units are millivolts (mV), microvolts (μV), and kilovolts (kV). The definition of voltage can be written as

$$v=\frac{\mathrm{d}w}{\mathrm{d}q} \qquad (1.1)$$

where v is the voltage measured in volts (V), w is the energy measured in joules (J), and q is the electric charge measured in coulombs (C).

As is shown in Fig. 1.1, the voltage between two nodes A and B v_{AB} equals to the energy required to move a unit charge from node B to node A.

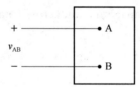

Fig. 1.1　A two-terminal element

A voltage is completely specified by both its magnitude and polarity. For the sake of simplicity, this book defines the reference polarity is a arbitrarily assigned polarity by a plus-minus sign pair.

1.1.3 Current

Current, also called current intensity, is a physical quantity that measures the speed at which a charge flows. Its magnitude is the amount of electricity passing through any cross section of the conductor per unit time. The direction of the current is defined as the direction of movement of the positive charge (proton), but it is difficult to know the real direction of movement of the positive charge from a macro perspective. Therefore, the reference direction of the current is defined as an arbitrary direction specified by an arrow. The unit of current is ampere (A), The commonly used units are milliamp (mA) and microamp (μA). The definition of current can be written as

$$i=\frac{\mathrm{d}q}{\mathrm{d}t} \qquad (1.2)$$

where i is the current measured in amperes (A), q is the charge measured in coulombs

Chapter 1 Circuit Elements and Circuit Variables

(C), and t is the time measured in seconds (s).

A current is completely specified by both its magnitude and direction. The moving direction of positive charge (proton) can be considered as the current's real direction, but it is hard to know the moving direction of positive charge from a macroscopical view. So, we define the reference direction that is the arbitrarily assigned direction by an arrow.

Q1.1 The current across an element is 2 A, calculate the charge entering the element during 1min.

Solution: The charge across the element during 1min is 2 A×60 s=120 C.

1.1.4 Power and energy

Power is the energy consumed or absorbed per unit time and is expressed as

$$P = \frac{dw}{dt} \quad (1.3)$$

where P is the power measured in watts (W), w is the energy measured in joules (J), and t is the time measured in seconds (s).

Before further discussion, we introduce a new concept, which is important to power calculation. The assignments of the reference polarity for voltage and the reference direction for current are entirely arbitrary. However, once you have assigned the references, you must follow the reference rule in the following analysis. The most widely used sign convention applied to these references is called passive sign convention.

Passive sign convention means that when the reference direction of the current drops across the element according to the reference of the polarity voltage, we use a positive sign in the voltage-current relationship (VCR) or power expressions; otherwise, we use a negative sign. We apply the passive sign convention in all the following analysis.

Recalling the definitions of voltage and current, Eq(1.3) can be rewritten as

$$P = \left(\frac{dw}{dq}\right)\left(\frac{dq}{dt}\right) = vi \quad (1.4)$$

Please note that Eq(1.4) need to follow the passive sign convention. If the passive sign convention is not applied, that is, if the reference direction for the current rise across the element, the power is

$$P = -vi \quad (1.5)$$

When P is positive, power is absorbed by the element, or power is being delivered to the element; when P is negative, the element supplies power, or power is being extracted from the element.

Energy is the accumulation of power over time, and power is the energy consumed or absorbed per unit time. Their relationship can be expressed as

$$w = \int_{-\infty}^{t} P dt \quad (1.6.a)$$

or

$$P = \frac{dw}{dt} \qquad (1.6.b)$$

Table 1.1 Common units of the basic circuit variables

Item	Common units
Voltage	Volt (V), Millivolt (mV=10^{-3} V), Microvolt (μV=10^{-6} V)
Current	Ampere (A), Milliampere (mA=10^{-3} A), Microampere (μA=10^{-6} A)
Power	Watt (W=V·A), Milliwatt (mW=10^{-3} W)

Q1.2 A 20V voltage drops across an element from terminal 2 to terminal 1 and a current of 4A enters terminal 2.

(1) Specify the value of v and i for the polarity references shown in Fig. 1.2.

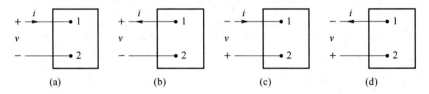

Fig. 1.2 The circuit of Q1.2(1)

(2) The circuit inside the box is absorbing or delivering power?

(3) How much power is absorbed by the circuit?

Solution:

(1) In subfigure (a) $v=-20$ V, $i=-4$ A; in subfigure (b) $v=-20$ V, $i=4$ A; in subfigure (c) $v=20$ V, $i=-4$ A; in subfigure (d) $v=20$ V, $i=4$ A.

(2) absorbing.

(3) 80W.

Q1.3 Calculate the power of the three subfigures in Fig. 1.3.

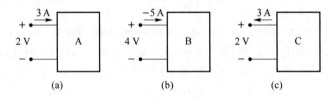

Fig. 1.3 The circuit of Q1.3

Solution:

(a) Following the passive sign convention, $P=(2\text{ V})(3\text{ A})=6$ W. Due to the power is positive, the power is absorbed by A or the power is being delivered to A.

(b) Following the passive sign convention, $P=(4\text{ V})(-5\text{ A})=-20$ W. Due to the power is negative, the power is supplied by B or the power is being extracted from B.

(c) Following the passive sign convention, $P=-(2\text{ V})(3\text{ A})=-6$ W. Due to the

Chapter 1 Circuit Elements and Circuit Variables

power is negative, the power is supplied by C or the power is being extracted from C.

1.2 Basic circuit element

In this subsection, we introduce some basic circuit elements, which are the important components in the circuits. In this book, all basic circuit elements can be considered as ideal basic circuit elements. An ideal two-terminal basic element has three attributes:

(1) It has only two terminals, which are points of connection to other circuit component;

(2) It is described mathematically in terms of current and/or voltage;

(3) It cannot be subdivided into other elements.

Here, the term "basic" means that such a circuit element cannot be further simplified or subdivided into other circuit elements; the word "ideal" means that such a circuit element does not exist in reality. But to simplify the analysis, all basic circuit elements in this book are considered as ideal basic circuit elements. Figure 1.4 is a representation of an ideal basic circuit element.

Fig. 1.4 An ideal two-terminal basic circuit element

1.2.1 Resistor

A resistor is a passive, two-terminal circuit element. In the circuit, resistor is mainly used to reduce the current, adjust the signal voltage, divide the voltage, and provide bias for the active component. Figure 1.5 is a resistor and its symbol in the circuit. A resistor is usually represented by a letter R. The magnitude of a resistor is expressed by the resistance value. For a resistor with uniform cross section, its resistance value is

$$R = \rho \frac{L}{A} \tag{1.7}$$

where R is the resistance value in ohms (Ω), ρ is the resistivity of the resistive material ($\Omega \cdot cm$), L is the length of the resistor (cm), and A is the cross-sectional area of the resistor (cm^2).

Fig. 1.5 Symbol and instance of resistor

The voltage-current relationship (VCR) of a resistor, also known as Ohm's Law, is that the voltage across a resistor is equal to the product of its resistance and the current flowing through it, and it needs to obey the passive sign convention. The VCR equation of a resistor is written as

$$v = iR \tag{1.8}$$

The power of a resistor can be expressed as

$$P = vi = i^2 R = \frac{v^2}{R} \tag{1.9}$$

1.2.2 Capacitor

A capacitor is a two-terminal passive electrical element, which can store energy in an electric field. It plays an important role on tuning circuit, bypassing circuit, coupling circuit, and filtering circuit, which can be applied in dynamic digital memories, high-pass or low-pass filters. In the circuit analysis, the capacitor current is zero if the voltage across the terminal is constant; and the capacitor current is infinite if the voltage across the terminal is time-varing. Fig. 1.6 is the symbol of the capacitor and some instances. The capacitor is usually denoted by the letter C in the circuit analysis.

Fig. 1.6 Symbol and instance of capacitor

A capacitor consists of two or more parallel conductive plates which are not connected or touching each other, but are electrically separated either by insulating material. When a voltage is applied between the two plates, it creates a potential difference and an electric field is established. In this condition, the capacitors store the electrical charges between its plates. By applying a voltage to a capacitor and measuring the charge on the plates, the ratio of the electric charge Q to the voltage v will give the capacitance value of the capacitor and is therefore given as

$$C = \frac{Q}{v} \tag{1.10}$$

where C is the capacitance measured in farads (F), Q is the electric charge measured in coulomb (C), v is the voltage measured in volts (V).

The current of a capacitor is proportional to the various rate of the voltage with the time, which is given by,

$$i = C \frac{dv}{dt} \tag{1.11}$$

Q1.4 The capacitance of a capacitor is 0.6 μF. The voltage of the capacitor is zero for $t < 0$ and is $40e^{-15\,000t} \sin 30\,000t$ V for $t \geq 0$.

(1) Calculate $i(0)$;

(2) Calculate the power of the capacitor when $t = \pi/80$ ms;

(3) Calculate the accumulated energy of the capacitor when $t = \pi/80$ ms.

Solution:

(1) $i = C\dfrac{dv}{dt} = (0.72\cos 30\,000t - 0.36\sin 30\,000t)e^{-15\,000t}$ A, $i(0) = 0.72$ A;

(2) $i(\pi/80 \text{ ms}) = -31.66$ mA, $v(\pi/80 \text{ ms}) = 20.505$ V, $p = vi = -649.23$ mW;

(3) $w = \int_{-\infty}^{t} p\,dt = \dfrac{1}{2}Cv^2 = 126.13\ \mu$J.

1.2.3 Inductor

An inductor is a passive two-terminal electrical element that stores energy in a magnetic field when electric current flows through it. An inductor typically consists of an insulated wire wound into a coil around a core. It is also a circuit element that prevents current fluctuations. The structure of an inductor is similar to a transformer, but with only one winding. Fig. 1.7 shows the symbol of the inductor and its actual instance. Inductive components are usually denoted by letter L in the circuit analysis. Its unit is Henry (H), the commonly used units are millihenry (mH) and microhenry (H).

The voltage of an inductor is proportional to the varing rate of the current with the time, which is given by,

$$v = L\frac{di}{dt} \tag{1.12}$$

where L is the inductance measured in henrys (H).

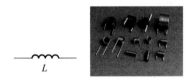

Fig. 1.7 Symbol and instance of inductor

Q1.5 As is shown in Fig. 1.8, the current of the inductor is zero for $t < 0$ and is $10te^{-5t}$ A for $t \geq 0$.

(1) Plot the waveform of the current;

(2) Calculate the voltage of the inductor;

(3) Plot the waveform of the voltage.

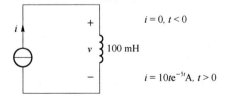

Fig. 1.8 The circuit of Q1.5

Solution:

(1) The waveform of the current is shown in Fig. 1.9.

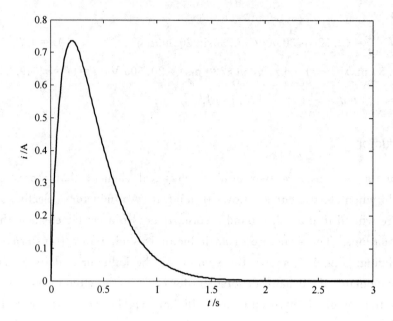

Fig. 1.9 The waveform of the current

(2) $v = L \dfrac{\mathrm{d}i}{\mathrm{d}t} = \mathrm{e}^{-5t}(1-5t)$ V for $t > 0$; $v = 0$ for $t < 0$.

(3) The waveform of the voltage is shown in Fig. 1.10.

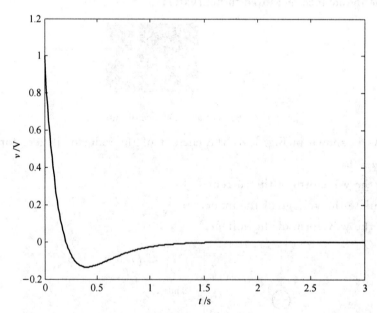

Fig. 1.10 The waveform of the voltage

Chapter 1 Circuit Elements and Circuit Variables

Table 1.2 Common units of basic circuit elements

Elements	Common units
Resistor	Ohm(Ω), Kiloohm(kΩ), Megaohm(MΩ)
Capacitor	Farads (F), Picofarad (pF), Microfarad (μF)
Inductor	Henry (H), Millihenry (mH), Microhenry(μH)

Table 1.3 Summary of VCR and power of resistor, capacitor and inductor

Elements	VCR	Power	Energy
Resistor	$v=iR$ (Ohm's Law)	$P=vi=i^2R=\dfrac{v^2}{R}$	$w=Pt=i^2Rt=\dfrac{v^2}{R}t$
Capacitor	$i=C\dfrac{dv}{dt}$	$P=vi$	$w=\displaystyle\int_{-\infty}^{t}P\,dt=\dfrac{1}{2}Cv^2$
Inductor	$v=L\dfrac{di}{dt}$	$P=vi$	$w=\displaystyle\int_{-\infty}^{t}P\,dt=\dfrac{1}{2}Li^2$

1.2.4 Independent voltage source and independent current source

An ideal independent voltage source is a circuit element that maintains a prescribed voltage across its terminals regardless of the value of the current flowing in those terminals. Fig. 1.11 (a) and (b) show the symbol and output characteristics of independent voltage source. Please note that both the polarity and magnitude of the voltage are provided for an independent voltage source.

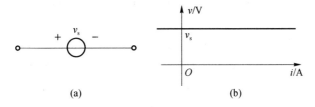

Fig. 1.11 Symbol and output characteristics of independent voltage source

An ideal independent current source is a circuit element that maintains a prescribed current through its terminals regardless of the voltage across those terminals. Fig. 1.12 (a) and (b) show the symbol and output characteristics of independent current source. Please note that both the direction and magnitude of the current are provided for an independent current source.

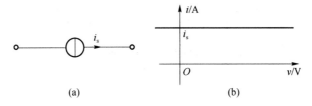

Fig. 1.12 Symbol and output characteristics of independent current source

1.2.5 Dependent source

An independent source establishes a voltage or current in a circuit without relying on voltage or current elsewhere in the circuit. The value of the voltage or current is specified by the value of the independent source alone. In the contrast, a dependent source establishes a voltage or current whose value depends on the value of a voltage or current elsewhere in the circuit. Dependent source is divided into four categories: voltage-control voltage source, current-control voltage source, voltage-control current source, and current-control current source, as is shown in Fig. 1.13.

To specify a voltage-control voltage source, you must identify the controlling voltage, the equation to compute the supplied voltage from the controlling voltage, and the reference polarity for the supplied voltage. In Fig. 1.13 (a), the supplied voltage v_s is written as

$$v_s = \mu v_x \tag{1.13}$$

where v_x is the controlling voltage, μ is voltage ratio factor (a dimensionless scaling factor).

Similarly, the supplied voltage v_s in current-control voltage source is

$$v_s = \rho i_x \tag{1.14}$$

where i_x is the controlling current and ρ is transfer resistance (a factor with unit V/A). The symbol of the current-control voltage source is shown in Fig. 1.13 (b).

The supplied current i_s in voltage-control current source is written as

$$i_s = \alpha v_x \tag{1.15}$$

where v_x is the controlling voltage and α is transfer conductance (a factor with unit A/V). The symbol of the voltage-control current source is shown in Fig. 1.13(c).

The supplied current i_s in current-control current source is written as

$$i_s = \beta i_x \tag{1.16}$$

where i_x is the controlling current and β is current ratio factor (a dimensionless scaling factor). The symbol of the current-control current source is shown in Fig. 1.13 (d).

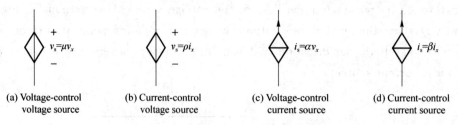

(a) Voltage-control voltage source   (b) Current-control voltage source   (c) Voltage-control current source   (d) Current-control current source

Fig. 1.13 Symbol of dependent source

… Chapter 1 Circuit Elements and Circuit Variables

1.3 Exercises

E1.1 The current entering an element is $i(t) = 2\cos 2\,000t$ A. Find the expression for $q(t)$.

E1.2 Two circuits, represented by boxes A and B, are connected as shown in Fig. 1.14. The reference direction for the current i in the interconnection and the reference polarity for the voltage v across the interconnection are shown in the figure. For each of the following sets of numerical values, calculate the power in the interconnection and state whether the power is flowing from A to B or vice versa.

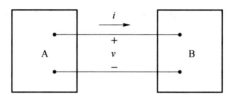

Fig. 1.14 The circuit of E1.2

E1.3 The voltage and current at the terminals of the circuit element are zero for $t<0$. For $t \geqslant 0$, they are $v = 5e^{-50t}\sin 150t$ V and $i = 2e^{-50t}\sin 150t$ A.

(1) Find the power absorbed by the element at $t = 10$ ms.

(2) Find the total energy absorbed by the element.

E1.4 As Fig. 1.15 shows, the current $i_s = 2$ A, calculate the voltage of the dependent source.

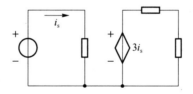

Fig. 1.15 The circuit of E1.4

E1.5 The voltage across a 20 μF capacitor is $v = 50\sin 200t$ V for $0 < t < 5\pi$ ms, calculate the charge, power and energy, sketch the energy waveform,

E1.6 The reference direction for the current and the reference polarity for the voltage are shown in Fig. 1.16. The voltage and current of element A are $v = -3$ V and $i = 5$ A respectively. The voltage and current of element B are $v = 5$ V and $i = -2$ mA respectively. Calculate how much power the two elements absorb?

Fig. 1.16 The circuit of E1.6

E1.7 In the circuit shown in Fig. 1.17, v_s denotes the voltage of the ideal voltage source. External circuit remains unchanged. Change the value of resistance R. Which branch current of the circuit will change?

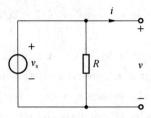

Fig. 1.17 The circuit of E1.7

E1.8 The $v-i$ characteristics of a branch containing a voltage source are shown in Fig. 1.18, in which the value of the resistance of different line is different. Try to analyze the relationship of value between R_1, R_2 and R_3.

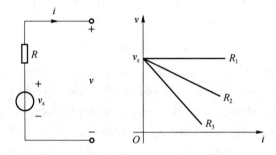

Fig. 1.18 The circuit of E1.8

E1.9 As Fig. 1.19 shows, state whether all elements are absorbing or supplying power for $v_s>0$, $i_s>0$.

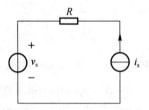

Fig. 1.19 The circuit of E1.9

E1.10 Verify the power conservation theorem in the circuit shown in Fig. 1.20.

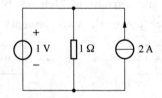

Fig. 1.20 The circuit of E1.10

E1.11 In the circuit shown in Fig. 1.21, the voltage source $v_s = 10$ V and the current source $i_s = 1$ A. Please state whether the voltage source and the current source are always supplying power to the resistance network.

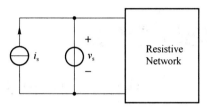

Fig. 1.21 The circuit of E1.11

Chapter 2

Basic Circuit Analysis Method

As mentioned in the forward, circuit analysis is mainly performed under the topology constraints and element constraints. This chapter will introduce the basic circuit analysis methods based on topology constraints.

The goal of Chapter 2: Basic circuit analysis method

(1) understand the definition of nodes, paths, loops, meshes and branches;

(2) skill in the equivalence of series and parallel combinations;

(3) understand and skill in Kirchhoff's Current Law (KCL) and Kirchhoff's Voltage Law (KVL);

(4) understand 2b-method;

(5) skill in node-voltage method;

(6) skill in mesh-current method.

2.1 Node, branch and mesh

In this book, all circuits are planar circuits, i. e., the circuits that can be draw on a plane with no crossing branches, as is shown in Fig. 2.1 (a). Under this background, some primary concepts of planar circuit are introduced in the following.

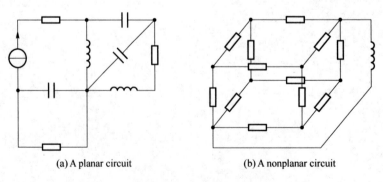

(a) A planar circuit   (b) A nonplanar circuit

Fig. 2.1   a planar circuit and a nonplanar circuit

1. Node

A point which two or more elements have a common connection is called a node. Note that the perfect conducting wires are considered as part of a node. Every element has a node at each of its ends.

Based on the definition of node, point A and H are one node, and point F, D and G are one node. Therefore, there are five nodes in Fig. 2.2.

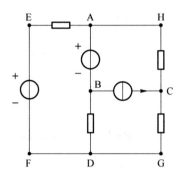

Fig. 2.2 An example of circuit model

2. Path, loop and branch

A path is a sequence of nodes with the property that each node in the sequence is adjacent to the node next to it.

A path in which last node is the same as the starting node, is called a loop. A loop is a closed path.

A branch is defined as a single path, composed of one simple element and the node at each end of it. Based on the definition of branch, the number of branches equals to that of elements. Therefore, there are seven branches in Fig. 2.2.

3. Mesh

A mesh is defined as a loop that does not contain any other loops within it. Meshes do not enclose any element. Based on the definition of mesh, there are three meshes in Fig. 2.3.

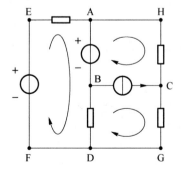

Fig. 2.3 An example of circuit model

Now, we try to find how many branches, nodes and meshes are in Fig. 2.4. The answer is 14 branches, 9 nodes and 6 meshes in Fig. 2.4 (a), and 10 branches, 7 nodes and 4 meshes in Fig. 2.4 (b).

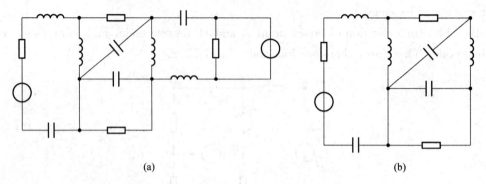

Fig. 2.4 Examples of circuit models

2.2 Kirchhoff's circuit law

Charge conservation and energy conservation are the basic laws of nature, and their expression in lumped circuits is Kirchhoff's circuit law. Kirchhoff's circuit law applies to DC circuits and AC circuits with circuit sizes much smaller than the wavelength of electromagnetic radiation. Kirchhoff's circuit law is divided into Kirchhoff's voltage law and Kirchhoff's current law, which respectively represent the constraint of voltage and current in the circuit. The two laws are described in detail below.

1. Kirchhoff's current law

Kirchhoff's current law (KCL) states that: at any time, at any node in the circuit, the algebraic sum of all currents entering and leaving a node is equal to zero, i.e., $\sum_{n=1}^{N} i_n(t) = 0$. In other words, the sum of the currents entering a node is equal to the sum of the currents leaving the node. Note that the reference direction of the current is critical for KCL. Although the reference direction of the current can be set arbitrarily, it is necessary to follow the set reference direction in subsequent analysis. There is an example to introduce KCL in Fig. 2.5.

Fig. 2.5 An example to introduce KCL

Based on the KCL, the KCL equation of Fig. 2.5 can be expressed as

$$i_A + i_B + (-i_C) + (-i_D) = 0 \qquad (2.1.a)$$

or
$$(-i_A)+(-i_B)+i_C+i_D=0 \tag{2.1.b}$$
or
$$i_A+i_B=i_C+i_D \tag{2.1.c}$$

Q2.1 Using KCL, try to calculate the currents in the circuit in Fig. 2.6.

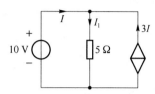

Fig. 2.6 The circuit of Q2.1

Solution: Using Ohm's Law, calculate the current I_1,
$$I_1=\frac{10\text{ V}}{5\ \Omega}=2\text{ A} \tag{2.2}$$

Based on KCL
$$I_1=I+3I \tag{2.3}$$

Then
$$I=0.5\text{ A} \tag{2.4}$$

2. Kirchhoff's voltage law

Kirchhoff's voltage law (KVL) states that: at any time, in any loop in the circuit, the algebraic sum of all voltages across a loop is zero, i.e., $\sum_{n=1}^{N} v_n(t) = 0$. Please note that both the reference direction and the reference polarity need to be considered in KVL. You can go around the loop in either clockwise or counterclockwise direction as the reference direction.

The algebraic sum of all the voltages around any loop in a circuit is zero, i.e., $\sum_{n=1}^{N} v_n(t) = 0$. KVL sign convention may be clockwise or anti-clockwise.

Based on the KVL, the KVL equation of Fig. 2.7 can be expressed as
$$-v_1+v_2-v_3=0 \tag{2.5.a}$$
or
$$v_1+v_3-v_2=0 \tag{2.5.b}$$

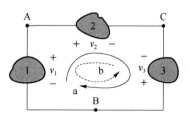

Fig. 2.7 An example to introduce KVL

Q2.2 Calculate the current I_0 in the circuit in Fig. 2.8 and calculate the power of each element in the circuit.

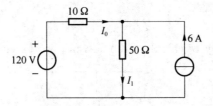

Fig. 2.8 The circuit of Q2.2

Solution: Based on KCL, we can obtain $I_1 - I_0 - 6\text{ A} = 0$.

Based on KVL, we can obtain $50 I_1 + 10 I_0 - 120\text{ }\Omega = 0$. Then, $I_0 = -3\text{ A}$ and $I_1 = 3\text{ A}$

The power of the current source is

$$P_I = -v_1(6\text{ A}) = -150\text{ V}(6\text{ A}) = -900\text{ W} \tag{2.6}$$

The power of the voltage source is

$$P_V = -120\text{ V}I_0 = 360\text{ W} \tag{2.7}$$

The power of the 50 Ω resistor is

$$P_{R=50\,\Omega} = 50\text{ }\Omega(3\text{ A})^2 = 450\text{ W} \tag{2.8}$$

The power of the 10 Ω resistor is

$$P_{R=10\,\Omega} = 10\text{ }\Omega(-3\text{ A})^2 = 90\text{ W} \tag{2.9}$$

Q2.3 Calculate the power of voltage source, current source and the resistor in Fig. 2.9.

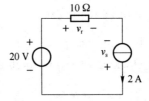

Fig. 2.9 The circuit of Q2.3

Solution: Using Ohm's Law and KVL, the voltage of the current source is

$$v_s = v_r - 20\text{ V} = 8\text{ V} - 20\text{ V} = -12\text{ V} \tag{2.10}$$

The power of the current source is

$$P_I = -(-12\text{ V}) \times 2\text{ A} = 24\text{ W} \tag{2.11}$$

The power of the voltage source is

$$P_V = -20\text{ V} \times 2\text{ A} = -40\text{ W} \tag{2.12}$$

The power of the resistor is

$$P_R = 8\text{ V} \times 2\text{ A} = 16\text{ W} \tag{2.13}$$

2.3 Series-parallel combinations of resistors

Elements in a circuit can be connected in a few different ways, the two simplest of which are series and parallel. In series connection, the current flows only through one path will be the same when passing through each resistor. In parallel connection, the voltage remains the same across each resistors in the circuit. In the following, we will analyze the series-parallel combination of resistance, capacitor and inductor.

1. Resistor

(1) The equivalent resistance of a set of resistors in series is shown in Fig. 2.10.

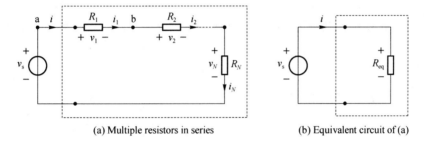

(a) Multiple resistors in series　　　　(b) Equivalent circuit of (a)

Fig. 2.10　Resistor in series

Based on KVL, we can obtain

$$v_s = v_1 + v_2 + \cdots + v_N \tag{2.14}$$

Then, given by Ohm's law, we further obtain

$$v_s = R_1 i_1 + R_2 i_2 + \cdots + R_N i_N \tag{2.15}$$

Due to the fact that all resistors are in series, the current is

$$i = i_1 = i_2 = \cdots = i_N \tag{2.16}$$

Then

$$v_s = (R_1 + R_2 + \cdots + R_N) i \tag{2.17}$$

If we assume $v_s = R_{eq} i$, we can obtain

$$R_{eq} = \sum_{n=1}^{N} R_n \tag{2.18}$$

i.e., the total resistance of resistors in series is equal to the sum of their individual resistances. The Fig. 2.10 (b) can be equivalent to Fig. 2.10 (a).

Resistors in series are common in circuits. The voltage divider circuit shown in Fig. 2.11 is a classic application example of resistors in series.

Based on KVL, Ohm's Law and the equivalence of resistors in series, the voltage of the two resistors are

$$\begin{cases} v_1 = \dfrac{R_1}{R_1 + R_2} v_s \\ v_2 = \dfrac{R_2}{R_1 + R_2} v_s \end{cases} \tag{2.19}$$

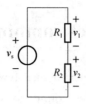

Fig. 2.11 A voltage divider circuit

Generally, for N resistors in series

$$v_n = \frac{R_n}{\sum_{i=1}^{N} R_i} v_s \tag{2.20}$$

(2) The equivalent resistance of a set of resistors in parallel

Based on KCL and Ohm's law, the current in Fig. 2.12 is given by

$$i_s = i_1 + i_2 + \cdots + i_N = \frac{v}{R_1} + \frac{v}{R_2} + \cdots + \frac{v}{R_N} = \frac{v}{R_{eq}} \tag{2.21}$$

If we assume $i_s = \frac{v}{R_{eq}}$, we can obtain

$$\frac{1}{R_{eq}} = \sum_{n=1}^{N} \frac{1}{R_n} \tag{2.22}$$

That is, the reciprocal of the equivalent resistance in the parallel circuit is equal to the sum of the reciprocals of the resistances of each resistor. The circuit in Fig. 2.12 (b) can be regarded as the equivalent circuit of the circuit in Fig. 2.12 (a).

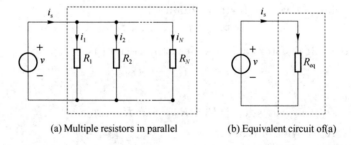

(a) Multiple resistors in parallel   (b) Equivalent circuit of(a)

Fig. 2.12 An example of resistors in parallel

In order to simplify Eq(2.22), a new concept is introduced—conductivity, which is expressed in letters G and the unit is Siemens (S). Conductivity is inversely proportional to resistance, i. e. $G_n = \frac{1}{R_n}$. Therefore, the equivalent conductivity of a resistor-parallel circuit can be written as

$$G_{eq} = \sum_{n=1}^{N} G_n \tag{2.23}$$

The equivalent conductivity of a resistor-series circuit can be written as

$$\frac{1}{G_{eq}} = \sum_{k=1}^{N} \frac{1}{G_k} \tag{2.24}$$

A two-parallel-resistor case is shown in Fig. 2.13. The equivalent resistance of two paralleled resistors is

$$R_{eq} = R_1 // R_2 = \frac{1}{\frac{1}{R_1} + \frac{1}{R_2}} = \frac{R_1 R_2}{R_1 + R_2} \tag{2.25}$$

Note that the equivalent resistance of three paralleled resistors is

$$R_{eq} = R_1 // R_2 // R_3 = \frac{R_1 R_2 R_3}{R_1 R_2 + R_1 R_3 + R_2 R_3} \tag{2.26}$$

instead of $\frac{R_1 R_2 R_3}{R_1 + R_2 + R_3}$

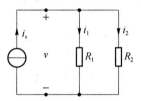

Fig. 2.13 Current divider circuit

Resistors in parallel are very common in circuits. The current divider circuit shown in Fig. 2.13 is a classic application of resistors in parallel. According to Ohm's Law and the equivalence of resistors in parallel, the voltage can be expressed as

$$v = i_1 R_1 = i_2 R_2 = \frac{R_1 R_2}{R_1 + R_2} i_s \tag{2.27}$$

Then we can obtain

$$\begin{cases} i_1 = \frac{R_2}{R_1 + R_2} i_s = \frac{G_1}{G_1 + G_2} i_s \\ i_2 = \frac{R_1}{R_1 + R_2} i_s = \frac{G_2}{G_1 + G_2} i_s \end{cases} \tag{2.28}$$

Generally, for N resistors in parallel,

$$i_n = \frac{G_n}{\sum_{i=1}^{N} G_i} i_s \tag{2.29}$$

Q2.4 Calculate the voltage v_o in Fig. 2.14.

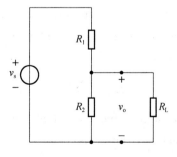

Fig. 2.14 The circuit of Q2.4

Solution: The circuit in Fig. 2.14 is neither in series nor in parallel. Instead, it included both the series part and parallel part. As Fig. 2.14 shows, R_2 and R_L are parallel and the corresponding equivalent resistance is given by

$$R_{eq} = \frac{R_2 R_L}{R_2 + R_L} \quad (2.30)$$

Then, like the voltage divider circuit, we can obtain

$$v_o = \frac{R_{eq}}{R_1 + R_{eq}} v_s \quad (2.31)$$

Q2.5 Find the equivalent resistance R_{ab} between terminal a and b in Fig. 2.15.

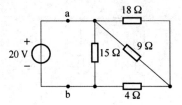

Fig. 2.15 The circuit of Q2.5

Solution: First, we replot the circuit in Fig 2.15 as that in Fig. 2.16 to enable the circuit topology clear.

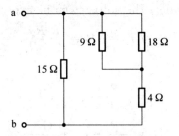

Fig. 2.16 The replot of Fig. 2.15

Then, based on the rules of series-parallel combinations of resistors, we obtain the equivalent resistance

$$R_{ab} = (9\ \Omega // 18\ \Omega + 4\ \Omega) // 15\ \Omega = (6\ \Omega + 4\ \Omega) // 15\ \Omega = 6\ \Omega \quad (2.32)$$

2. Inductor

The series-parallel combination of inductors also follows the same rules as resistor, that is, the equivalent inductance of all uncoupled inductors in series is equal to the sum of their respective inductances, i.e.,

$$L_{eq} = \sum_{n=1}^{N} L_n \quad (2.33)$$

and the reciprocal of the equivalent inductance of all uncoupled inductors in parallel is equal to the sum of their respective reciprocals, i.e.,

$$\frac{1}{L_{eq}} = \sum_{n=1}^{N} \frac{1}{L_n} \quad (2.34)$$

The two results in Eq (2.33) and Eq (2.34) can be proved by KCL and KVL. The proof in detail is provided after learning VCR of an inductor in Chapter 4, only the conclusion is given here.

3. Capacitor

The series-parallel combination of capacitors is exactly different from the series-parallel combination of inductors, that is, the inverse of the equivalent capacitance of all capacitors in series is equal to the sum of their respective reciprocals, i.e.,

$$\frac{1}{C_{eq}} = \sum_{n=1}^{N} \frac{1}{C_n} \tag{2.35}$$

and the equivalent capacitance of capacitors in parallel is equal to the sum of their individual capacitances, i.e.,

$$C_{eq} = \sum_{n=1}^{N} C_n \tag{2.36}$$

The two results in Eq (2.35) and Eq(2.36) can be proved by KCL and KVL. The proof in detail is provided after learning VCR of a capacitor in Chapter 4, only the conclusion is given here.

Thus, the equivalent combinations of circuit elements is an effective method to simplify the circuit analysis.

2.4 2b-method

In this subsection, a simple and general circuit analysis method, i.e., 2b-method, is present, which is based on KCL, KVL and Ohm's Law. Two other circuit analysis methods, i.e., node voltage method and mesh current method, are introduced in the following two subsections.

First, we talk about the meaning of 2b. Now, we focus on a circuit with n nodes and b branches. Based on KCL, we can write $(n-1)$ independent equations for any $(n-1)$ nodes; Based on KVL, we can write $m = b - (n-1)$ independent equations for m meshes; Based on Ohm's Law, we can write b VCR equations for b branches. Thus, there are $(n-1)$ KCL equations, $b-(n-1)$ KVL equations, and b VCR (Ohm's Law) equations. Summing all these equations up, we can obtain $2b$ independent equations to find $2b$ unknowns that are the current and voltage to be determined for $2b$ elements in the circuit. That is why this method is called 2b-method. Please note that if there are s independent sources in a circuit, the unknown variables in the circuit is $s + 2(b-s) = (2b-s)$.

Q2.6 Using 2b-method to calculate all currents and voltages in the circuit, as is shown in Fig. 2.17.

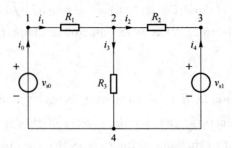

Fig. 2.17 The circuit of Q2.6

Solution: There are $n=4$ nodes and $b=5$ branches in Fig. 2.17. So, we can write $n-1=3$ KCL equations for 3 nodes (e.g., node 1,2,3).

$$\begin{cases} i_0 - i_1 = 0 \\ i_1 - i_2 - i_3 = 0 \\ i_2 + i_4 = 0 \end{cases} \quad (2.37)$$

The KCL equation for node 4 (i.e., $-i_0 + i_3 - i_4 = 0$) can be deduced from the equations of node 1,2,3. Therefore, it is not an independent equation.

Apply KVL to $b-(n-1)=2$ meshes (e.g., mesh I and II in Fig. 2.18),

$$\begin{cases} v_1 + v_3 - v_{s0} = 0 \\ -v_3 + v_2 + v_{s1} = 0 \end{cases} \quad (2.38)$$

Moreover, for $b=5$ branches, the VCR equations are

$$\begin{cases} v_1 = R_1 i_1 \\ v_2 = R_2 i_2 \\ v_3 = R_3 i_3 \\ v_{s0} = \text{given} \\ v_{s1} = \text{given} \end{cases} \quad (2.39)$$

By integrating the 10 independent equations listed above, all current and voltage values in the circuit can be obtained.

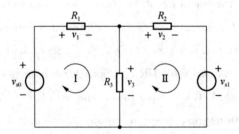

Fig. 2.18 Mesh of the circuit of Fig. 2.17

The procedure of $2b$-method is summarized as:

Step 1. Build $2b$ independent equations for a given circuit involving $(n-1)$ KCL equations for $(n-1)$ nodes, $b-(n-1)$ KVL equations for $b-(n-1)$ meshes, and b VCR (Ohm's Law) equations for b branches.

Step 2. Solve equations to determine currents and voltages for all (or required) circuit elements.

2.5 Node voltage method

In the previous analysis, we give a general method to calculate every unknown circuit variables. However, in some cases, it is not necessary to solve all variables, only some of them are our interest. Therefore, how to quickly find the required variables are the focus in the following two subsections.

In this and the next section, two circuit analysis methods for quickly solving the designated variables will be introduced, namely the node voltage method and the mesh current method. This section first introduces the node voltage method.

The node voltage method is an analytical method that uses the node voltage as a variable to write KCL equations. The node voltage is the voltage rise relative to the reference node voltage (commonly is set as zero).

Now we introduce the procedure of node voltage method.

Step 1: Select reference node and define node voltages.

Step 2: Build KCL equations for nodes except the reference node.

Step 3: Solve equations to get node voltages.

Step 4: Determine required unknowns by solved node voltages.

Q2.7 Use node voltage method to find v_1, v_2, and i_1 in Fig. 2.19.

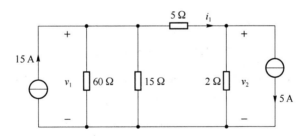

Fig. 2.19 The circuit of Q2.7

Solution: We use node voltage method to solve this problem.

Step 1: Select reference node and define node voltages as is shown in Fig. 2.20.

Step 2: Build KCL equations for nodes except the reference node

$$\begin{cases} 15\ \text{A} - \dfrac{v_a}{60\ \Omega} - \dfrac{v_a}{15\ \Omega} - \dfrac{v_a - v_b}{5\ \Omega} = 0 \\ \dfrac{v_a - v_b}{5\ \Omega} - \dfrac{v_b}{2\ \Omega} - 5\ \text{A} = 0 \end{cases} \qquad (2.40)$$

Step 3: Solve equations to get node voltages

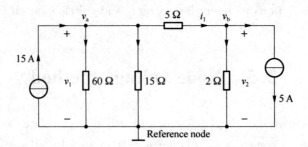

Fig. 2.20 The circuit of Fig. 2.19 with reference and non-reference nodes

$$\begin{cases} v_a = 60 \text{ V} \\ v_b = 10 \text{ V} \end{cases} \tag{2.41}$$

Step 4: Determine required unknowns by solved node voltages, $v_1 = v_a = 60$ V, $v_2 = v_b = 10$ V and $i_1 = \dfrac{v_a - v_b}{5} = \dfrac{60 \text{ V} - 10 \text{ V}}{5 \text{ }\Omega} = 10$ A.

Q2.8 Use node voltage method to find v_1, v_2, v_3 and v_4 in Fig. 2.21.

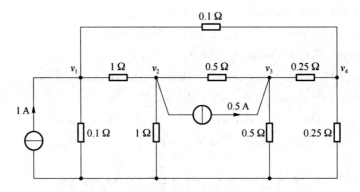

Fig. 2.21 The circuit of Q2.8

Solution: Select reference node and build KCL equations for other nodes

$$\begin{cases} 1 - 0.1v_1 - 0.1(v_1 - v_4) - (v_1 - v_2) = 0 \\ (v_1 - v_2) - v_2 - 0.5(v_2 - v_3) - 0.5 = 0 \\ 0.5(v_2 - v_3) + 0.5 - 0.5v_3 - 0.25(v_3 - v_4) = 0 \\ 0.1(v_1 - v_4) + 0.25(v_3 - v_4) - 0.25v_4 = 0 \end{cases} \tag{2.42}$$

Solve equations to obtain the node voltage

$$\begin{cases} v_1 = 1.23 \text{ V} \\ v_2 = 0.42 \text{ V} \\ v_3 = 0.67 \text{ V} \\ v_4 = 0.48 \text{ V} \end{cases} \tag{2.43}$$

Please note that if there are voltage source (independent and dependent voltage source) in the branches, the current flowing through it cannot be represented by the node voltage. Therefore, the KCL equations of two nodes at both ends of the voltage source branch cannot be written. In this time, we need to deal with this problem according to the following two conditions.

(1) If one node in the voltage source branch is the reference node, the node voltage of the other node is known (for independent voltage source) or can be expressed out (for dependent voltage source), so it is not necessary to build KCL equation for the node;

(2) If both ends of voltage source are not the reference node, two equations are needed. One is the KCL equation for the super node, and the other is the voltage difference equation for the voltage source branch.

A new concept appears. What is the super node?

If the voltage source and two nodes at both ends of this branch are combined into one node, then the node is a super node. Each supernode contains two nodes, one is a non-reference node, and the other can be either a non-reference node or a reference node. According to the definition of super node, node 1, node 4 and the current-controlled voltage source in Fig. 2.21 constitute a super node. According to the definition of KCL, KCL is also true for super nodes.

Q2.9 Use node voltage method to find the value of v_3 and v_4.

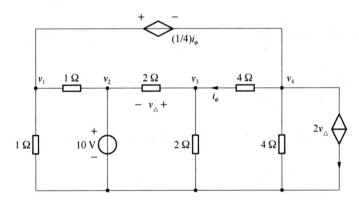

Fig. 2.22 The circuit of Q2.9

Solution: Select the reference node. Then, we build KCL equations for nodes, and write KCL and voltage difference equation for supernode.

$$\begin{cases} 2(v_2-v_3)-2v_3+4(v_4-v_3)=0 \text{------KCL\_eqation\_for\_node3} \\ v_2=10 \text{------eqution\_for\_node2} \\ -v_1+(v_2-v_1)-4(v_4-v_3)-4v_4-2(v_3-v_2)=0 \text{-----KCL\_equation\_for\_Super\_Node} \\ v_1-v_4=\frac{1}{4}\times 4(v_4-v_3) \text{------equation\_for\_voltage\_difference} \end{cases}$$

(2.44)

Solve equations to obtain the node voltage $v_3=4.5$ V and $v_4=4$ V.

Q2.10 The circuit is shown in Fig. 2.23. The power of R_2 is 2 W, find the resistance value of R_1, R_2 and R_3.

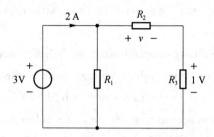

Fig. 2.23 The circuit of Q2.10

Solution: Using KVL to calculate the voltage of R_2,

$$v_2 = 3\text{ V} - 1\text{ V} = 2\text{ V} \tag{2.45}$$

Then we can derive R_2 reversely,

$$R_2 = \frac{v^2}{P} = \frac{(2\text{ V})^2}{2\text{ V}\cdot\text{A}} = 2\text{ }\Omega \tag{2.46}$$

Using Ohm's law, the current across R_2 and R_3 are

$$I_2 = \frac{v}{R_2} = \frac{2\text{ V}}{2\text{ }\Omega} = 1\text{ A} \tag{2.47}$$

$$R_3 = \frac{1}{I_2} = \frac{1\text{ V}}{1\text{ A}} = 1\text{ }\Omega \tag{2.48}$$

Using KCL, the current across R_1 is given by

$$I_1 = 2\text{ A} - I_2 = 2\text{ A} - 1\text{ A} = 1\text{ A} \tag{2.49}$$

Then $R_1 = \dfrac{3\text{ V}}{I_1} = \dfrac{3\text{ V}}{1\text{ A}} = 3\text{ }\Omega$.

2.6 Mesh current method

In this subsection, we introduce the second alternative method, called mesh current method. The mesh current method is a circuit analysis method that uses mesh current (not branch current) to formulate circuit equations. The currents before this subsection are branch currents. The branch current is the actual current through the circuit, and the mesh current is a virtual current flowing along the mesh boundary. In the circuit analysis, it is assumed that there is only one mesh current per mesh. Fig. 2.24 shows the relationship between mesh current and branch current.

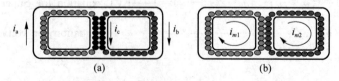

Fig. 2.24 The relationship between mesh current and branch current

In Fig. 2.24 (a) the current i_a, i_b and i_c are the branch currents while the currents i_{m1}

and i_{m2} in Fig. 2.24 (b) are the mesh currents. The relationship between them are $i_a = i_{m1}$, $i_b = i_{m2}$ and $i_c = i_{m1} - i_{m2}$.

Now we introduce the procedure of mesh current method.

Step 1. Mark mesh current and its reference direction (e. g. , clockwise) for every mesh;

Step 2. Build KVL equations for every mesh;

Step 3. Solve equations for the mesh currents;

Step 4. Determine the required unknown currents and voltages by the mesh currents.

Q2.11 In Fig. 2.25, $R_1 = 5$ Ω, $R_2 = 10$ Ω, $R_3 = 20$ Ω, $v_{s0} = 20$ V, $v_{s1} = 10$ V. Determine the currents of every branch.

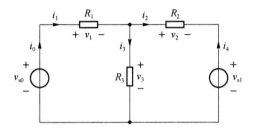

Fig. 2.25 The circuit of Q2.11

Solution: Mark mesh current i_a and i_b as is shown in Fig. 2.26.

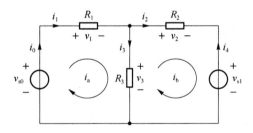

Fig. 2.26 The circuit of Fig. 2.25 with mesh currents

Build KVL equations for all meshes

$$\begin{cases} R_1 i_a + R_3 (i_a - i_b) - v_{s0} = 0 \\ R_2 i_b + v_{s1} + R_3 (i_b - i_a) = 0 \end{cases} \quad (2.50)$$

Solve the equations to get mesh current i_a and i_b

$$i_a = \frac{8}{7} \text{ A}, \quad i_b = \frac{3}{7} \text{ A} \quad (2.51)$$

Determine the branch currents

$$i_0 = i_1 = i_a = \frac{8}{7} \text{ A}, \quad i_2 = i_b = \frac{3}{7} \text{ A}, \quad i_3 = i_a - i_b = \frac{5}{7} \text{ A and } i_4 = -i_b = -\frac{3}{7} \text{ A}$$

Please note that if there are some current sources (independent and dependent current sources) in the branches, their voltages cannot be expressed by mesh currents. Therefore,

KVL equations for the meshes consisting of current source branches cannot be built. In this time, we need to deal with this problem according to the following two conditions.

(1) If the current source branch is at the edge of the circuit, this mesh current is known (for independent current sources) or can be expressed out (for dependent current sources), so it is not necessary to build KVL equation for the mesh current;

(2) If the current source branch is not at the edge of the circuit, there are two meshes across the current source branch, thereby two equations should be built. One is KVL equation for the super mesh, and the other is the current different equation for the current source branch.

A new concept appears-what is the super mesh?

A supermesh is a larger loop which has both meshes inside. Supermesh occurs when a current source is shared by two meshes. This leads to a KVL equation that involves two mesh currents. The other equation represents that the current of the current source is equal to one of the mesh currents minus the other.

Q2.12 Use mesh current method to find the total power dissipated in the circuit in Fig. 2.27.

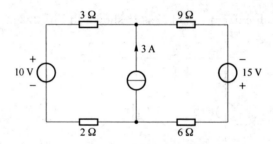

Fig. 2.27 The circuit of Q2.12

Solution: Mark mesh current i_a and i_b in Fig. 2.28 and the supermesh is labeled by dotted line.

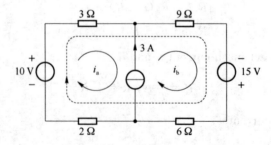

Fig. 2.28 The circuit of Fig. 2.27 with mesh currents

Build KVL equation and current difference equation

$$\begin{cases} i_b - i_a = 3 \text{ A} \\ i_a(3 \text{ }\Omega + 2 \text{ }\Omega) + i_b(9 \text{ }\Omega + 6 \text{ }\Omega) - 15 \text{ V} - 10 \text{ V} = 0 \text{ V} \end{cases} \qquad (2.52)$$

So $i_a = -1$ A and $i_b = 2$ A.

The power delivered to the resistors are $i_a^2(3 \text{ }\Omega + 2 \text{ }\Omega) + i_b^2(9 \text{ }\Omega + 6 \text{ }\Omega) = 65$ W and the power extracted to the three sources is $-10i_a - 15i_b + 3(15 - 6i_b - 9i_b) = -65$ W.

Base on the analysis of node voltage method and mesh current method. We compare the two methods in the following

(1) Mesh current analysis only applies to planar circuits;

(2) Node voltage method obtain the voltage directly, whereas mesh current analysis provides mesh currents (instead of branch currents);

(3) Mesh current analysis can provide $m = b - (n-1)$ equations where $m = b - (n-1) < b < 2b$; node voltage analysis supports $(n-1)$ equations where $n - 1 \leqslant b < 2b$. Therefore, both the node voltage method and the mesh current can reduce the number of equations to be solved and simplify the circuit analysis process.

2.7 Exercises

E2.1 How many branches, nodes, meshes exist in Fig. 2.29?

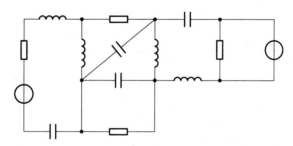

Fig. 2.29 The circuit of E2.1

E2.2 Calculate the equivalent resistance in Fig. 2.30 (a) and (b).

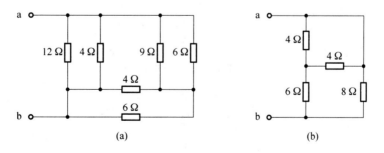

Fig. 2.30 The circuit of E2.2

E2.3 Calculate the equivalent resistance in Fig. 2.31.

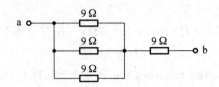

Fig. 2.31 The circuit of E2.3

E2.4 Find v_1, v_2 and i_1 in Fig. 2.32.

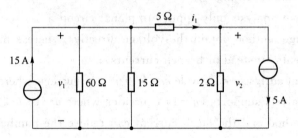

Fig. 2.32 The circuit of E2.4

E2.5 Given the circuit shown in Fig. 2.33, find (a) the value of i_a, i_b and v_o, (b) the power dissipated in each resistor and the power delivered by the voltage source.

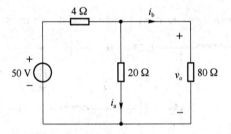

Fig. 2.33 The circuit of E2.5

E2.6 The current i_o in the circuit in Fig. 2.34 is 4 A. (a) Find i_1, (b) find the power dissipated in each resistor, (c) verify that the total power dissipated in the circuit equals the power supported by the 180 V source.

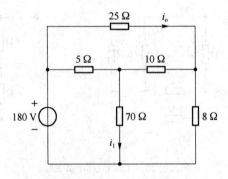

Fig. 2.34 The circuit of E2.6

E2.7 Find i_2, i_1 and i_o in the circuit in Fig. 2.35.

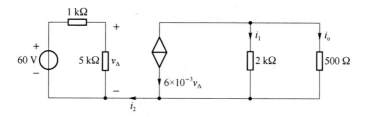

Fig. 2.35 The circuit of E2.7

E2.8 Find v_1 and v_2 in the circuit shown in Fig. 2.36.

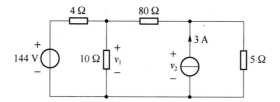

Fig. 2.36 The circuit of E2.8

E2.9 Find v_o in the circuit shown in Fig. 2.37.

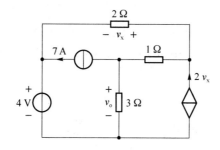

Fig. 2.37 The circuit of E2.9

E2.10 Find the total power developed in the circuit in Fig. 2.38 and check your answer by showing that the total power developed equals the total power dissipated.

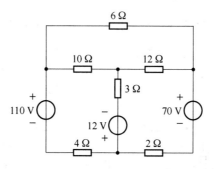

Fig. 2.38 The circuit of E2.10

E2.11 Solve i_Δ for in the circuit in Fig. 2.39. Find the power delived by the independent current source. Find the power delivered by the dependent voltage source.

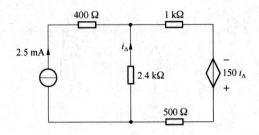

Fig. 2.39　The circuit of E2.11

E2.12　Find the total power developed in the circuit in Fig. 2.40.

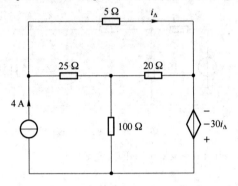

Fig. 2.40　The circuit of E2.12

E2.13　In Fig. 2.41, the voltage of the resister R is $v=8$ V, find R.

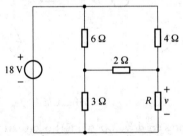

Fig. 2.41　The circuit of E2.13

E2.14　Use the mesh current method to find the branch current $i_1 \sim i_6$ in the circuit in Fig. 2.42.

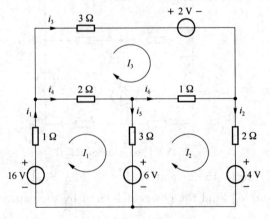

Fig. 2.42　The circuit of E2.14

E2.15 Use the mesh current method to find the current I in the circuit in Fig. 2.43.

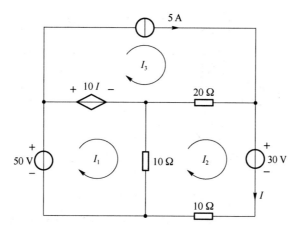

Fig. 2.43 The circuit of E2.15

E2.16 Use the mesh current method to find the voltage v in the circuit in Fig. 2.44.

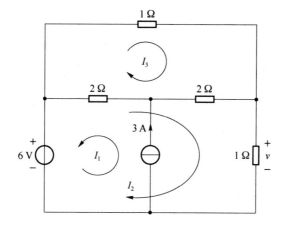

Fig. 2.44 The circuit of E2.16

E2.17 In Fig. 2.45, $v_{ab}=5$ V, find the voltage v_s in the circuit.

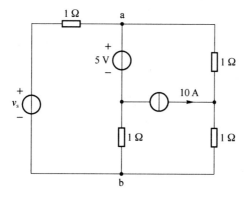

Fig. 2.45 The circuit of E2.17

E2.18 Use node voltage method to find the voltage v_{ab} in Fig. 2.46.

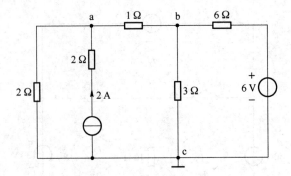

Fig. 2.46 The circuit of E2.18

E2.19 Use node voltage method to find the current i_1 and i_2 in Fig. 2.47.

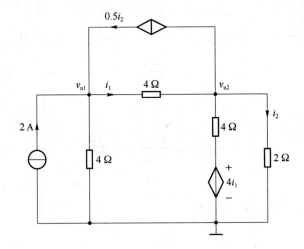

Fig. 2.47 The circuit of E2.19

E2.20 Find i_1, i_2 and i_3 in the circuit shown in Fig. 2.48.

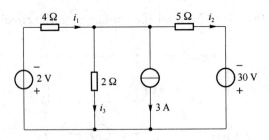

Fig. 2.48 The circuit of E2.20

Chapter 3

Circuit Theorem

Chapter 2 introduced the basic analysis methods in circuit theory. These methods can directly analyze simple circuits. However, as the circuit topology becomes more complicated, it is hoped that a simplified method can be found to split, combine or equivalent complex circuits. Therefore, this chapter will introduce several basic circuit theorems for simplifying complex circuits.

The goal of Chapter 3: Basic circuit theorem

(1) understand and skill in the superposition theorem;

(2) understand and skill in source transformation;

(3) understand and sill in Thevenin theorem and Norton theorem;

(4) understand and sill in the maximum power transfer.

3.1 Superposition theorem

If a linear circuit is excited by more than one independent source, the total response is the algebraic sum of the individual responses. An individual response is the result of an independent source acting alone. Acting alone means that each independent source is considered one time, meanwhile, other independent sources are "killed", or "turned off", or "zeroed out".

Noticeably, (1) if a voltage source is zeroed out, it is treated as a short circuit; (2) if a current source is zeroed out, it is treated as an open circuit; (3) superposition theorem does not apply to power calculation.

Q3.1 Use superposition theorem to calculate current i_1 and i_2.

Solution: Two independent sources are in the circuit in Fig. 3.1. According to superposition theorem, the circuit can be divided as two sub-circuits, as is shown in Fig. 3.2 (a) and (b) respectively. In the following, we analyze the two sub-circuits independently.

For the circuit in Fig. 3.2 (a)

$$i_1^{(1)} = i_2^{(1)} = \frac{54 \text{ V}}{9 \text{ }\Omega + 8 \text{ }\Omega} = \frac{54 \text{ V}}{27 \text{ }\Omega} = 2 \text{ A} \tag{3.1}$$

For the circuit in Fig. 3.2 (b)

$$i_1^{(2)} = -\frac{18\ \Omega}{9\ \Omega + 18\ \Omega} \times 6\ \text{A} = -4\ \text{A} \tag{3.2}$$

$$i_2^{(2)} = \frac{9\ \Omega}{9\ \Omega + 18\ \Omega} \times 6\ \text{A} = 2\ \text{A} \tag{3.3}$$

Then, based on the superposition theorem, the currents are the algebraic sum of the results of the two sub-circuits

$$i_1 = i_1^{(1)} + i_1^{(2)} = 2\ \text{A} - 4\ \text{A} = -2\ \text{A} \tag{3.4}$$

$$i_2 = i_2^{(1)} + i_2^{(2)} = 2\ \text{A} + 2\ \text{A} = 4\ \text{A} \tag{3.5}$$

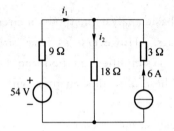

Fig. 3.1   The circuit of Q3.1

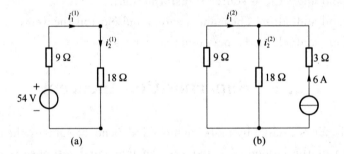

Fig. 3.2   Two sub-circuits of Fig. 3.1

Q3.2   Use superposition theorem to calculate voltage v in Fig. 3.3.

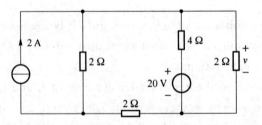

Fig. 3.3   The circuit of Q3.2

Solution: Two independent sources are in the circuit in Fig. 3.3. According to superposition theorem, the circuit can be divided as two sub-circuits as is shown in Fig. 3.4 (a) and (b). In the following, we can analyze the two sub-circuits independently.

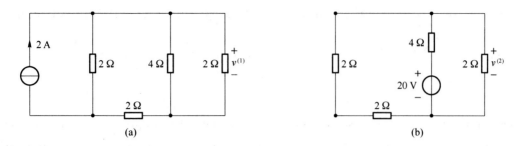

Fig. 3.4 Two sub-circuits of Fig. 3.3

We can obtain the voltages in the two sub-circuits $v^{(1)} = 1$ V and $v^{(2)} = 5$ V. Therefore,

$$v = v^{(1)} + v^{(2)} = 1 \text{ V} + 5 \text{ V} = 6 \text{ V} \tag{3.6}$$

3.2 Source transformations

In the circuit, a source is like a black box. The relationship between the voltage of the source and the current across the source describes all the properties of the source. If two sources have the same VCR, they can be considered as the same source.

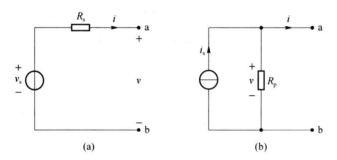

Fig. 3.5 Source transformation

As shown in Fig. 3.5 (a), the voltage source can be represented as an element providing a fixed voltage in series with a resistor. At this time, the voltage provided by the voltage source to the external circuit is

$$v = v_s - iR_s \tag{3.7}$$

where v_s is the open circuit voltage and R_s is the resistor in series with the voltage source.

As shown in Fig. 3.5 (b), the current source can be represented as an element providing a fixed current in parallel with a resistor. At this time, the current provided by the current source to the external circuit is

$$i = i_s - \frac{v}{R_p} \tag{3.8.a}$$

or

$$v = i_s R_p - iR_p \tag{3.8.b}$$

where i_s is the short circuit current and R_p is the resistor in parallel with the current source.

Observing Eq (3.7) and Eq (3.8), we found that if

$$\begin{cases} R_s = R_p \\ v_s = i_s R_p \end{cases} \qquad (3.9)$$

Then the voltage and current sources in Fig. 3.5 (a) and (b) can provide the same voltage and current for external circuits, that is, the voltage source and the current source are equivalent.

Source transformation allows a voltage source in series with a resistor to be replaced by a current source in parallel with the same resistor, or vice versa, as is written in Eq (3.9)

Note that

(1) The equivalent relationship between the voltage source and the current source is only for the external circuit, and is not equivalent to the inside of the source.

(2) During the equivalent conversion, the reference directions of the two sources are shown in Fig. 3.6 (a) and (b).

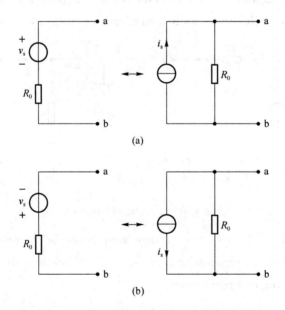

Fig. 3.6   The reference directions in source transformation

An extra question is what will happen if there is a resistor in parallel with the voltage source or a resistor in series with the current source, as is shown in Fig. 3.7 (a) and (b) respectively. In both cases, the resistor has no effect on the equivalent circuit with respect to terminal a and b.

Source transformation is not limited to a resistor circuit. By representing circuit elements as impedance in the phasor domain, source transformation can also be applied to a circuit that includes a capacitor and an inductor.

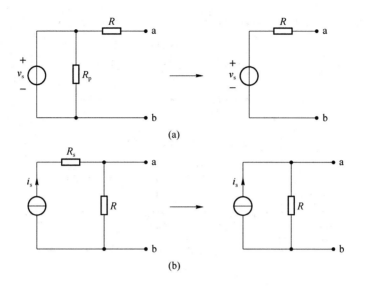

Fig. 3.7 Special case in source transformation

Q3.3 Use source transformation to calculate the current i in Fig. 3.8.

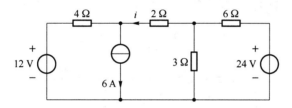

Fig. 3.8 The circuit of Q3.3

Solution: Based on the source transformation, we can replot Fig. 3.8 as Fig. 3.9.

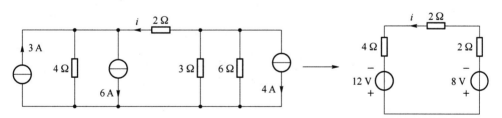

Fig. 3.9 The equivalent circuit of Fig. 3.8

So the current is given by

$$i = \frac{12 \text{ V} - 8 \text{ V}}{4 \text{ }\Omega + 2 \text{ }\Omega + 2 \text{ }\Omega} = \frac{4 \text{ V}}{8 \text{ }\Omega} = 0.5 \text{ A} \qquad (3.10)$$

Q3.4 Find the current in the 5 kΩ resistor by source transformation.

Solution: Using the source transformation, we can obtain the following transformation from Fig. 3.11 (a) to (g) orderly.

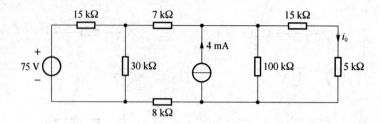

Fig. 3.10　The circuit of Q3.4

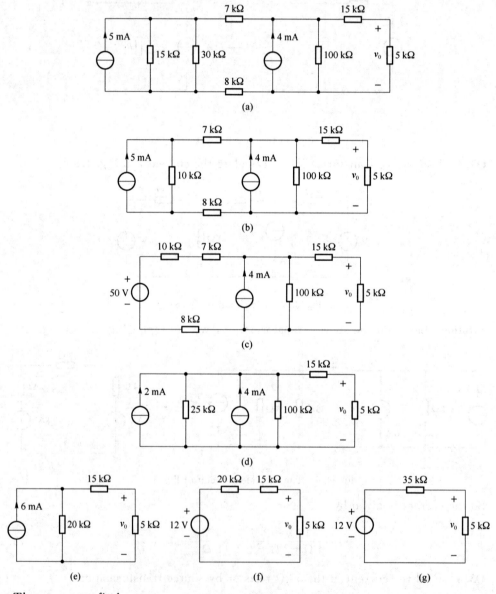

Fig. 3.11　The equivalent circuit of Fig. 3.10

Then, we can find

$$i_0 = \frac{12 \text{ V}}{35 \text{ k}\Omega + 5 \text{ k}\Omega} = 3 \text{ mA}$$

3.3 Thevenin theorem and Norton theorem

Sometimes, in circuit analysis, we want to concentrate on what happens at a specific pair of terminals. For example, when we plug a toaster into an outlet, we are interested primarily in the voltage and current at the terminal of the toaster. We have little or no interest in the effect that connecting the toaster has on voltages or currents elsewhere in the circuit supplying the outlet. That is to say, we want to focus on the behavior of the circuit supplying the outlet, but only at the outlet terminals. Thevenin theorem and Norton theorem are circuit simplification techniques that focus on terminal behavior and thus are extremely valuable aids in analysis.

In Chapter 2, the equivalence of series-parallel combinations has been introduced, but the actual circuits are hybrid connected together by resistors and sources. In addition to the step-by-step series-parallel conversion, is there any other simpler equivalent method?

Thevenin theorem and Norton theorem are circuit simplification techniques used to analyze one-port network in circuits.

We first describe Thevenin theorem. Thevenin theorem, also referred as equivalent voltage source theorem, states that a linear network with terminal a and b can be equivalent to a resistance in series with an independent voltage source, where the voltage value v_{Th} of the voltage source is equal to the open circuit voltage v_{oc} when terminal a and b are open, and the resistance R_{Th} of the resistor is equal to the equivalent resistance when all independent sources in the network are set to zero.

As shown in Fig. 3.12, it is a Thevenin equivalent circuit of a network with terminal a and b. Regardless of the topology and parameters of network N, as long as the circuits can be equivalent to the same Thevenin equivalent circuit, they provide the same voltage and current for the external circuit at terminal a and b. When they connect to the same load, the voltage and current on the load are same.

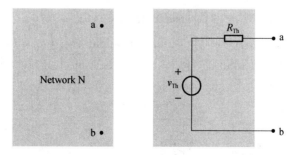

Fig. 3.12 An example of terminals a and b network

Q3.5 Find Thevenin equivalent circuit between terminals a and b.

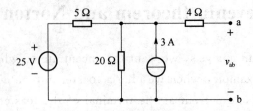

Fig. 3.13 The circuit of Q3.5

Solution: Find the open circuit voltage between two terminals a and b.

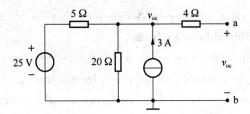

Fig. 3.14 The circuit of Q3.5 when terminals a and b is open-circuit

By node voltage analysis method, we have

$$\frac{v_{oc}-25}{5}+\frac{v_{oc}}{20}-3=0 \qquad (3.10)$$

So the open circuit voltage is $v_{oc}=32$ V and $v_{Th}=v_{oc}=32$ V.

Now we find Thevenin equivalent resistance between terminal a and b by deleting all sources in the circuit

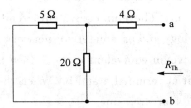

Fig. 3.15 The circuit of Q3.5 without any source

According to Fig. 3.15, the Thevenin equivalent resistance $R_{Th}=8$ Ω. Then, the Thevenin equivalent of the circuit between terminals a and b is shown in Fig. 3.16.

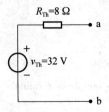

Fig. 3.16 Thevenin equivalent circuit for Q3.5

Based on Thevenin equivalent circuit shown in Fig. 3.12, and source transformation described in the previous subsection, a voltage source in series with a resistance can be equivalent to a current source in parallel with a resistance, that is, Fig. 3.12 can be equivalent to Fig. 3.17. The current value of the current source and the resistance of the parallel resistor can also be calculated according to Eq (3.9).

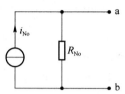

Fig. 3.17 Norton equivalent circuit

Norton's theorem, also known as the equivalent current source theorem, state that: a linear network with terminal a and b can be equivalent to a resistance in parallel with an independent current source, where the current value i_{No} of the current source is equal to the short circuit current i_{sc} when terminal a and b directly connect, and the resistance R_{No} of the resistor is equal to the equivalent resistance when all independent sources in the network are set to zero. In addition, source transformation tell us that

$$\begin{cases} R_{No} = R_{Th} \\ i_{sc} = \dfrac{v_{Th}}{R_{Th}} \end{cases} \tag{3.11}$$

Similar to Thevenin's theorem, regardless of the topology and parameters of network N, as long as the circuits can be equivalent to the same Norton equivalent circuit, they provide the same voltage and current for the external circuit at terminal a and b. When they connect to the same load, the voltage and current on the load are same.

By using Thevenin theorem or Norton theorem, the complex one-port linear resistive network (excluding the load part) can be simplified. The application of the two theorems in AC circuit is exactly the same as in DC circuit, but the resistance is extended to impedance.

3.4 Maximum power transfer theorem

In practical applications, we are concerned about whether the circuit can provide enough power for the load to support its normal operation. That is to say, we want to provide as much power as possible to the load. Therefore, this section discusses the relationship between load and power transfer in the circuit.

Based on the analysis in the last subsection, any one-port linear resistive circuit with the independent source can be represented as the circuit model in Fig. 3.18, where R_L is the load between the terminal a and b. The power transferred to the load R_L is given by

$$P = \left(\frac{v_{Th}}{R_{Th} + R_L}\right)^2 R_L \tag{3.12}$$

Let $\dfrac{dP}{dR_L} = 0$, we can obtain the maximum power delivered to R_L, i.e.,

$$P_{max} = i^2 R_L = \left(\frac{v_{Th}}{R_{Th} + R_L}\right)^2 R_L = \frac{v_{Th}^2}{4R_L} \quad \text{for} \quad R_{Th} = R_L \tag{3.13}$$

The above results are recorded as the maximum power transfer theorem. The theorem reveals the load value needed to obtain the maximum power when the circuit is determined. That is, when the load resistance is equal to the internal resistance of the power supply circuit, the load can obtain the maximum power.

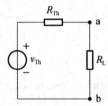

Fig. 3.18 Maximum power transfer circuit

If the resistance of the load is greater than the resistance of the power supply, the power transferred from the power supply to the load will be higher, but the power of the load will be reduced due to the increase of the total resistance in the circuit. If the resistance of the load is less than the resistance of the power supply, although the total resistance is reduced, most of the power will be consumed in the power supply. At this time, the power obtained by the load will be reduced.

The theorem tells us how to choose load resistance to realize the maximum power transfer when the power supply resistance is given. However, the theorem does not explain how to choose the source resistance for a given load resistance! People often have a misunderstanding about this theorem, that is, choosing the same resistance value for the power supply as the load value can maximize the load power. In fact this conclusion is totally wrong, Because the source resistance that maximizes load power is always zero, regardless of the value of the load resistance.

The maximum powert ransfer theorem can also be extended to AC circuits, that is, when the load impedance is equal to the complex conjugate of the source impedance, the maximum power transfer can be realized. The proof of this conclusion will be given in Chapter 5.

3.5 Exercises

E3.1 Use superposition theorem to calculate the voltage v in Fig. 3.19.

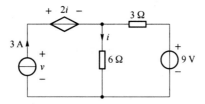

Fig. 3.19 The circuit of E3.1

E3.2 Use superposition theorem to calculate the current i in Fig. 3.20.

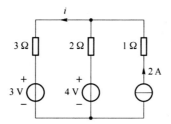

Fig. 3.20 The circuit of E3.2

E3.3 In Fig. 3.21, $v_s = 20$ V.
(a) Calculate the value of i;
(b) Calculate v_s if $i = 0.25$ A.

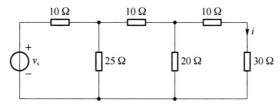

Fig. 3.21 The circuit of E3.3

E3.4 Use superposition theorem to calculate the current i in Fig. 3.22.

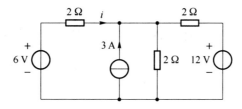

Fig. 3.22 The circuit of E3.4

E3.5 Use source transformation to calculate i_1 and i_2 in Fig. 3.23, and then calculate the power of 18 Ω resistor and the power of the voltage source.

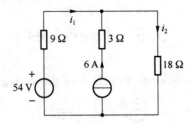

Fig. 3.23 The circuit of E3.5

E3.6 Use source transformation to calculate the current i in Fig. 3.24.

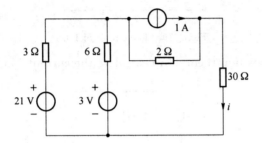

Fig. 3.24 The circuit of E3.6

E3.7 Use source transformation to calculate the voltage v_{ab} in Fig. 3.25.

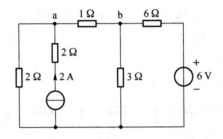

Fig. 3.25 The circuit of E3.7

E3.8 Use a series of source transformations to find the current i_0 in the circuit in Fig. 3.26.

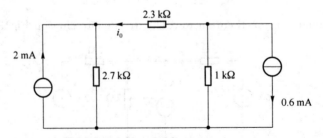

Fig. 3.26 The circuit of E3.8

E3.9   Use a series of source transformations to find the current i_0 in the circuit in Fig. 3.27.

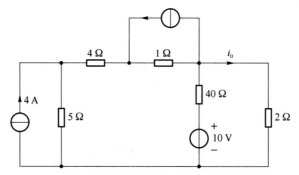

Fig. 3.27   The circuit of E3.9

E3.10   Use the principle of superposition to find the voltage v_0 in the circuit in Fig. 3.28.

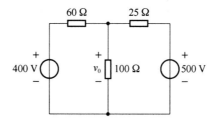

Fig. 3.28   The circuit of E3.10

E3.11   Use the principle of superposition to find the voltage v_0 in the circuit in Fig. 3.29.

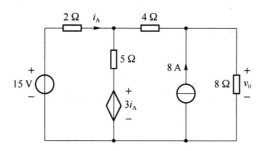

Fig. 3.29   The circuit of E3.11

E3.12   Find the Thevenin equivalent with respect to the terminal a and b for the circuit in Fig. 3.30.

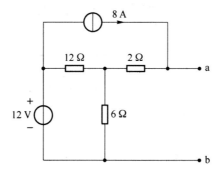

Fig. 3.30   The circuit of E3.12

E3.13 Find the Thevenin equivalent with respect to the terminal a and b for the circuit in Fig. 3.31.

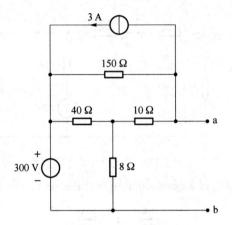

Fig. 3.31 The circuit of E3.13

E3.14 Find the Norton equivalent with respect to the terminal a and b for the circuit in Fig. 3.32.

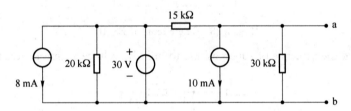

Fig. 3.32 The circuit of E3.14

E3.15 Find the Thevenin equivalent with respect to the terminal a and b for the circuit in Fig. 3.33.

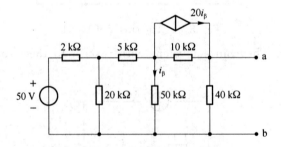

Fig. 3.33 The circuit of E3.15

E3.16 Find the Thevenin equivalent with respect to the terminal a and b for the circuit in Fig. 3.34.

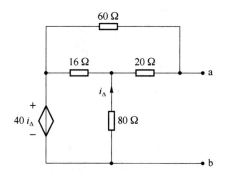

Fig. 3.34 The circuit of E3.16

E3.17 The variable resistor in the circuit in Fig. 3.35 is adjusted for maximum power transfer to R_0.

(a) Find the value of R_0.

(b) Find the maximum power that can be delivered to R_0.

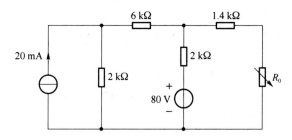

Fig. 3.35 The circuit of E3.17

E3.18 Find the Thevenin equivalent with respect to the terminal a and b for the circuit in Fig. 3.36.

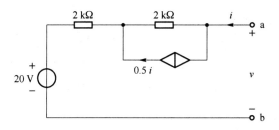

Fig. 3.36 The circuit of E3.18

E3.19 Find the Norton equivalent with respect to the terminal a and b for the circuit in Fig. 3.37.

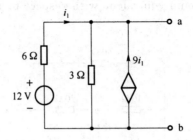

Fig. 3.37　The circuit of E3.19

E3.20　Shown in Fig. 3.38, N is a linear source-resistance network. While switch S1 and S2 are open, the voltmeter reads 6V. While S1 is closed and S2 is open, the voltmeter reads 4V. Find the voltmeter reading while S1 is open and S2 is closed.

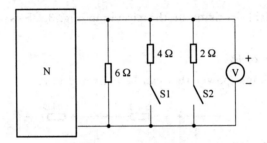

Fig. 3.38　The circuit of E3.20

E3.21　Fine the resistance value of R_L at which the maximum power is obtained and the value of the maximum power.

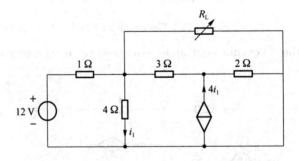

Fig. 3.39　The circuit of E3.21

Chapter 4

RC Circuit and RL Circuit

In the previous chapters, we learned the basic methods and basic theorems in circuit analysis, and used this knowledge to analyze some simple circuits that only composed of resistors and voltage or current sources. we know that resistors are static elements. In this chapter, the circuits containing dynamic elements (e. g. , inductors and capacitors) will be analyzed.

The goal of Chapter 4: First-order circuit

(1) understand the characteristics of inductor and capacitor, skill in their VCR and the equivalence of series and parallel combinations;

(2) understand switching rules;

(3) understand the definition of first-order circuit, zero-input and zero-state response of RL and RC circuit;

(4) skill in calculating zero-input response of RL and RC circuit;

(5) skill in calculating zero-state response of RL and RC circuit;

(6) understand instantaneous and steady response, and skill in calculating the complete response and three-factor method.

4.1 First-order circuits

In Chapter 1, we have learned that an inductor is an element that can convert electrical energy into magnetic energy and then store it. Its VCR is expressed as

$$v = L \frac{\mathrm{d}i}{\mathrm{d}t} \tag{4.1}$$

i. e. , the voltage of an inductor is equal to the change of the current multiplied by the inductance, and meanwhile the voltage and current obey the passive sign convention. When the rate of change of the current is zero, the voltage is zero. In this condition, the inductor acts like a short circuit.

If the voltage is bounded, the current is

$$i(t) = \frac{1}{L}\int_{-\infty}^{t} v(\zeta)\mathrm{d}\zeta = \frac{1}{L}\int_{-\infty}^{t_0} v(\zeta)\mathrm{d}\zeta + \frac{1}{L}\int_{t_0}^{t} v(\zeta)\mathrm{d}\zeta = i(t_0) + \frac{1}{L}\int_{t_0}^{t} v(\zeta)\mathrm{d}\zeta \tag{4.2}$$

It can be known from Eq (4.2) that the current at a certain time is not only related to the voltage at this moment, but also related to all voltage values before this time, which indicates that the inductor current has memory. When the inductor voltage is bounded, the inductor current is continuous function. This is a very important for circuit analysis with inductor.

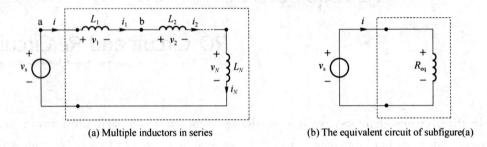

(a) Multiple inductors in series   (b) The equivalent circuit of subfigure(a)

Fig. 4.1   Inductors in series

There are N inductors in series Fig. 4.1 (a). According to KVL, we obtain

$$v_s = v_1 + v_2 + \cdots + v_N \tag{4.3}$$

Based on VCR, then

$$v_s = L_1 \frac{di_1}{dt} + L_2 \frac{di_2}{dt} + \cdots + L_N \frac{di_N}{dt} \tag{4.4}$$

Because all inductors are in series, the currents

$$i = i_1 = i_2 = \cdots = i_N \tag{4.5}$$

Then

$$v_s = (L_1 + L_2 + \cdots + L_N) \frac{di}{dt} \tag{4.6}$$

If we assume $v_s = L_{eq} \frac{di}{dt}$, then

$$L_{eq} = \sum_{n=1}^{N} L_n \tag{4.7}$$

That is, the equivalent inductance of all inductors in series is equal to the sum of their respective inductances. The circuit in Fig. 4.1 (b) can be regarded as the equivalent circuit of the circuit in Fig. 4.1 (a).

Based on KVL and VCR of inductor, the current in Fig. 4.2 is given by

$$i_s = i_1 + i_2 + \cdots + i_N = \int \frac{v}{L_1} dt + \int \frac{v}{L_2} dt + \cdots + \int \frac{v}{L_N} dt = \left(\frac{1}{L_1} + \frac{1}{L_2} + \cdots + \frac{1}{L_N} \right) \int v \, dt \tag{4.8}$$

If we assume $i_s = \int \frac{v}{L_{eq}} dt$, then

$$\frac{1}{L_{eq}} = \sum_{n=1}^{N} \frac{1}{L_n} \tag{4.9}$$

That is, the reciprocal of the equivalent inductance in the parallel circuit is equal to the sum of the reciprocals of these inductances. The circuit in Fig. 4.2 (b) can be regarded as the equivalent circuit of the circuit in Fig. 4.2 (a).

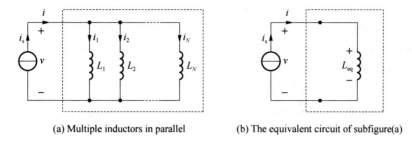

(a) Multiple inductors in parallel   (b) The equivalent circuit of subfigure(a)

Fig. 4.2   Inductors in parallel

4.2   Capacitor

In Chapter 1, we have learned that a capacitor is a passive double-ended circuit element that stores energy in an electric field. Its VCR is expressed as

$$i = C\frac{dv}{dt} \tag{4.10}$$

That is, the current "across" a capacitor is equal to the change of the voltage multiplied by the capacitance. Meanwhile the voltage and current obey the passive sign convention. When the voltage change rate is zero, the current is zero. In this condition, a capacitor is equivalent to an open circuit.

If the current is bounded, the voltage is

$$v(t) = \frac{1}{C}\int_{-\infty}^{t} i(\zeta)d\zeta = \frac{1}{C}\int_{-\infty}^{t_0} i(\zeta)d\zeta + \frac{1}{C}\int_{t_0}^{t} i(\zeta)d\zeta = v(t_0) + \frac{1}{C}\int_{t_0}^{t} i(\zeta)d\zeta \tag{4.11}$$

It can be known from Eq (4.11) that the voltage at a certain time is not only related to the current at that time, but also related to all current values before this time, which indicates that the capacitor voltage has memory.

When the capacitor current is bounded, the capacitor voltage is a continuous function. This is a very important for circuit analysis with capacitors.

There are N capacitors in series Fig. 4.3 (a). According to KVL, we obtain

$$v_s = v_1 + v_2 + \cdots + v_N \tag{4.12}$$

Based on VCR, then

$$v_s = \int \frac{i_1}{C_1}dt + \int \frac{i_2}{C_2}dt + \cdots + \int \frac{i_N}{C_N}dt \tag{4.13}$$

Because all capacitors are in series, the currents

$$i = i_1 = i_2 = \cdots = i_N \tag{4.14}$$

Then

$$v_s = \left(\frac{1}{C_1} + \frac{1}{C_2} + \cdots + \frac{1}{C_N}\right)\int i\,dt \tag{4.15}$$

If we assume $v_s = \frac{1}{C_{eq}}\int i\,dt$, then

$$\frac{1}{C_{eq}} = \sum_{n=1}^{N} \frac{1}{C_n} \tag{4.16}$$

That is, the reciprocal of the equivalent capacitance in the series circuit is equal to the sum of the reciprocal of these capacitances. The circuit in Fig. 4.3 (b) can be regarded as the equivalent circuit of the circuit in Fig. 4.3 (a).

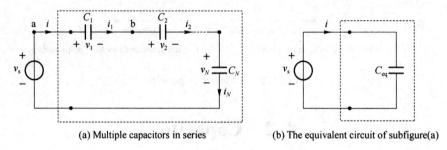

(a) Multiple capacitors in series (b) The equivalent circuit of subfigure(a)

Fig. 4.3 Capacitors in series

Based on KVL and VCR of inductor, the current in Fig. 4.4 is given by

$$i_s = i_1 + i_2 + \cdots + i_N = C_1 \frac{dv}{dt} + C_2 \frac{dv}{dt} + \cdots + C_N \frac{dv}{dt} \tag{4.17}$$

If we assume $i_s = C_{eq} \frac{dv}{dt}$, then

$$C_{eq} = \sum_{n=1}^{N} C_n \tag{4.18}$$

That is, the total capacitance of all paralleled capacitors is equal to the sum of their respective capacitances. The circuit in Fig. 4.4 (b) can be regarded as the equivalent circuit of the circuit in Fig. 4.4 (a).

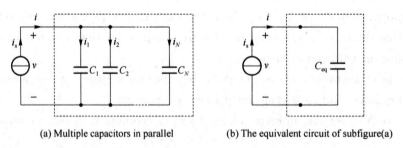

(a) Multiple capacitors in parallel (b) The equivalent circuit of subfigure(a)

Fig. 4.4 Capacitors in parallel

4.3 Switching rule

The circuit usually works in a stable state, at which time the voltage and current in the circuit are fixed values or change periodically. When a circuit changes from one working state to another (e. g. , open or close a switch, circuit parameters change, etc.), it is referred to circuit switching. Changing the circuit will change the voltage and current in the circuit and this change will last for a period of time. This process of change is called transient state.

If there are some memory elements such as capacitors or inductors in the circuit, instantaneous change cannot occur due to their memorability. Therefore, the capacitor voltage and inductor current before and after the switching remain same.

$$v_C(t_0^+) = v_C(t_0^-) \tag{4.19}$$

$$i_L(t_0^+) = i_L(t_0^-) \tag{4.20}$$

where $v_C(t_0^-)$ and $i_L(t_0^-)$ are the steady state of the capacitor voltage and inductor current before the switching.

Eq (4.19) and Eq (4.20) represent the switching rule. The capacitor voltage and the inductor current follow the switching rule, from which the initial values after switching can be derived, and on this basis, the initial values after switching of other variables are derived.

4.4 First-order circuit

The previous chapters mainly focus on the analysis and discussion of the circuit composed of resistors and DC sources. This chapter begins to focus on the circuit composed of DC sources, resistors, and dynamic elements such as inductors or capacitors.

A circuit with a dynamic element is called a first-order circuit, which is usually represented by a first-order differential equation. If multiple inductors or capacitors in the original circuit can be represented by an equivalent inductor or capacitor, the circuit can also be regarded as a first-order circuit. Commonly, first-order circuits include *RL* (resistor-inductor) circuit and *RC* (resistor-capacitor) circuit.

The analysis of the first-order circuit can be divided into the following three cases.

(1) The current and voltage generate in a circuit when the energy stored in an inductor or capacitor is suddenly released into a resistive network. This happens when the inductor or capacitor is abruptly disconnected from the DC source. In this case, the circuit can be simplified into two equivalent forms as shown in Fig. 4.5. The response caused by the stored energy inside the dynamic (energy storage) element in the circuit is called zero-input response of the circuit. Zero input response emphasizes that the nature of the circuit itself, not external sources of excitation, determines its behavior.

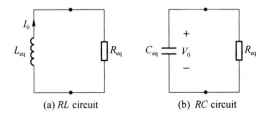

(a) *RL* circuit (b) *RC* circuit

Fig. 4.5 The zero-input response of *RL* circuit and *RC* circuit

(2) In the second condition, we consider the currents and voltages that arise when energy is being acquired by an inductor or capacitor due to the sudden application of a DC voltage or current source. In this case, the circuit can be simplified into two equivalent forms as shown in Fig. 4.6. The currents and voltages that arise in this configuration are referred as the zero-state response of the circuit, to emphasize that the effect of the excitation from the external sources, not the initial storing energy in the dynamic elements (inductor or capacitor).

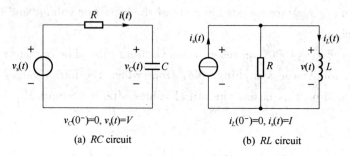

Fig. 4.6 The zero-state response of RC circuit and RL circuit

(3) In the third condition, both the initial storing energy in dynamic elements and the excitation from the external sources are considered. Accordingly, we call the response under this condition is complete response. The complete response can be regarded as the sum of the zero-input and zero-state response. In this case, the circuit can be simplified into two equivalent forms as shown in Fig. 4.7.

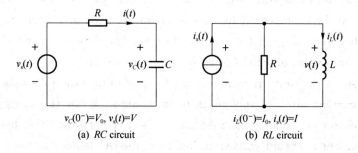

Fig. 4.7 The complete response of RC circuit and RL circuit

4.5 Zero-input response of *RL* and *RC* circuit

In the following, we discuss the zero-input response of RC and RL circuits, respectively.

4.5.1 Zero-input response of *RC* circuit

Fig. 4.8 shows an RC circuit with zero-input response. The circuit is composed of a DC voltage source V_0, resistor R_1 and R, a capacitor C and a switch k. The switch k has been in the position a for a long time, until the switch k is moved from position a to b at time $t=0$. Then find $v_C(t)$ for $t>0$.

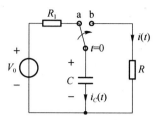

Fig. 4.8 The zero-input response of a RC circuit

The capacitor can be consider as an open circuit when the source is a DC voltage source. After a long time, all voltages and currents have reached a steady state.

We use $t=0^-$ to represent the moment before switching and $t=0^+$ to represent the initial moment after switching. Due to the continuity of capacitor voltage, no sudden change will occur. Therefore, the voltage at both ends of the capacitor is constant at the moment after switching from position a to position b, then the initial value of the capacitor voltage after switching is

$$v_C(0^+)=v_C(0^-)=V_0 \tag{4.21}$$

The switch k is at position b for $t>0$. According to KVL equation,

$$v_C(t)=i(t)R \tag{4.22}$$

and VCR of a capacitor is

$$i(t)=-i_C(t)=-C\frac{dv_C(t)}{dt} \tag{4.23}$$

A first-order homogeneous differential equation can be obtained from (4.22) and (4.23)

$$v_C(t)+RC\frac{dv_C(t)}{dt}=0 \tag{4.24}$$

Combined with the initial value of capacitor voltage shown in (4.21), the solution of the equation is

$$v_C(t)=V_0 e^{-\frac{t}{RC}} \tag{4.25}$$

where V_0 is the initial capacitor voltage after switching. In addition, the change of capacitor voltage depends on the value of RC.

Now we introduce a new concept—time constant. In Eq (4.25), $\frac{1}{RC}$ determines the rate at which the current or voltage approaches to zero. The reciprocal of this ratio is defined as the time constant of the circuit, denote by τ. The time constant of a RC circuit equals to the product of the resistor and capacitor, namely,

$$\tau=RC \tag{4.26}$$

From Eq(4.26), we can deduce that the time constant RC governs the rate of decay.

As shown in Fig. 4.9, the zero-input response curve of capacitor voltage in an RC circuit is depicted.

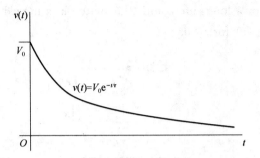

Fig. 4.9 The zero-input response of a RC circuit

Thus, the analysis of zero-input response of an RC circuit can be divided into three steps

(1) Calculate the initial value of the capacitor voltage $v_C(0^+)=v_C(0^-)$;

(2) Write the first order homogeneous differential equation according to KVL equation of the circuit and VCR equation of the capacitor;

(3) Find the time constant $\tau=RC$;

(4) Solve the equations, and obtain the capacitor voltage $v_C(t)=v_C(0^+)e^{-\frac{t}{\tau}}$.

4.5.2 Zero-input response of RL circuit

Fig. 4.10 shows the zero-input response of an RL circuit. The circuit is composed of the DC current source I_0, the resistor R_0 and R, the inductor L and a switch k. The switch k has been closed for a long time, until it opens at time $t=0$. Then find $i_L(t)$ for $t>0$.

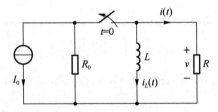

Fig. 4.10 A circuit used to illustrate the zero-input response of a RL circuit

The switch is closed and the current source is in parallel with resistor R_0, R and inductor L for $t<0$. After a period of time, the inductor can be regarded as short circuit in the DC source cicuit, and the current flowing through the inductor is equal to the current from the DC current source. The circuit reaches a stable state.

We use $t=0^-$ to represent the moment before the switch is turned on and use $t=0^+$ to represent the initial moment after the switch is turned on. Due to the continuity of the inductive current, no change will occur. Therefore, the inductor current remain the same at the moment when the switch is opened, then the initial value of the inductor current after opening the switch is given by

$$i_L(0^+)=i_L(0^-)=I_0 \tag{4.27}$$

The switch k is open for $t>0$. According to KVL equation,
$$v_L(t)=i(t)R \tag{4.28}$$
and VCR of an inductor
$$v_L(t)=L\frac{di_L}{dt}=-L\frac{di}{dt} \tag{4.29}$$
A first-order homogeneous differential equation can be obtained from Eq (4.28) and Eq (4.29)
$$i_L(t)+\frac{L}{R}\frac{di_L(t)}{dt}=0 \tag{4.30}$$
Combined with the initial value of inductor current shown in Eq (4.27), the solution of the equation is
$$i_L(t)=I_0 e^{-\frac{Rt}{L}} \tag{4.31}$$
where I_0 is the initial inductor current after the switching. In addition, the change of inductor current depends on the value of $\frac{L}{R}$.

In Eq(4.31), $\frac{L}{R}$ determines the rate at which the current or voltage approaches to zero. The reciprocal of this ratio is the time constant of the circuit, denote by $\tau=\frac{L}{R}$. From Eq(4.31), we can deduce that the time constant $\tau=\frac{L}{R}$ governs the rate of decay.

As shown in Fig. 4.11, the zero-input response curve of inductor current in an RL circuit is depicted.

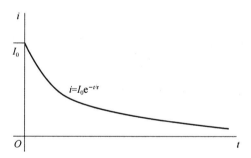

Fig. 4.11 The zero-input response of a RL circuit

Thus, the analysis of zero-input response of an RL circuit can be divided into three steps

(1) Calculate the initial value of the inductor current $i_L(0^+)=i_L(0^-)=I_0$;

(2) Write the first order homogeneous differential equation according to KVL equation of the circuit and VCR equation of the inductor;

(3) Find the time constant $\tau=\frac{L}{R}$;

(4) Solve the equations, and obtain the inductor current $i_L(t)=i_L(0^+)e^{-\frac{t}{\tau}}$.

4.6 Zero-state response of *RL* and *RC* circuit

We consider the currents and voltages that arise when energy from a DC voltage or current source is being acquired by an inductor or capacitor. Thus, we simplify the circuit to one of the two equivalent forms shown in Fig. 4.2. The currents and voltages that arise in this configuration are referred as the zero-state response of the circuit, to emphasize that the effect of the excitation from the external sources, not the initial storing energy in the dynamic elements.

As shown in Fig. 4.12, the step function can be used to describe the DC power supply. For example, a step function $v(t)$ can be expressed as

$$v(t) = \begin{cases} V_0, & t \geq t_0 \\ 0, & t < t_0 \end{cases} \quad (4.32)$$

This section analyzes the zero-state response of *RC* and *RL* circuit.

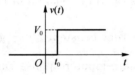

Fig. 4.12 A step function

4.6.1 Zero-state response of *RL* circuit

Fig. 4.13 is an *RL* circuit. The circuit is composed of the DC voltage source $v_s(t)\big|_{t \geq 0} = V_s$, the resistor R, the inductor L and a switch k. The switch k is closed at time $t = 0$. Then find $i_L(t)$ for $t > 0$.

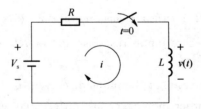

Fig. 4.13 A circuit used to illustrate the zero-state response of *RL* circuit.

Before the switch k is closed, the current in the *RL* circuit is zero, i.e., $i_L(0^-) = 0$. After the switch k has been closed, a first order homogeneous differential equation according to KVL equation of the circuit and VCR equation of the inductor is given by

$$Ri(t)+L\frac{di(t)}{dt}=v_s(t) \tag{4.33}$$

The solution to the differential equation in the form $a_1 y'(t)+a_0 y(t)=f(t)$ with the initial condition $y(0)=Y_0$ is $y(t)=y_h(t)+y_p(t)$, where $y_h(t)$ is the solution to the homogeneous equation $a_1 y'(t)+a_0 y(t)=0$ and $y_p(t)$ is the particular solution to the equation $a_0 y(t)=f(t)$.

First, the solution to the homogeneous equation $Ri(t)+L\frac{di(t)}{dt}=0$ is

$$i_h(t)=Ae^{-\frac{t}{\tau}} \tag{4.34}$$

where $\tau=\frac{L}{R}$ and A is a coefficient need to be decided

Then the particular solution is

$$i_p(t)=\frac{V_s}{R} \tag{4.35}$$

So, the solution is $i(t)=i_h(t)+i_p(t)=Ae^{-\frac{t}{\tau}}+\frac{V_s}{R}$.

Combing the initial condition, we obtain $i(0)=A+\frac{V_s}{R}=0$, i.e., $A=-\frac{V_s}{R}$. Thus, the zero-state response of the RC circuit in Fig. 4.14 is

$$i(t)=\frac{V_s}{R}(1-e^{-\frac{t}{\tau}}) \tag{4.36}$$

where the term $\frac{V_s}{R}$ in Eq(4.36) is the steady value of the inductor current $i_L(\infty)=\frac{V_s}{R}$. From Eq(4.36), we can deduce that the time constant governs the rate of grouth. Fig. 4.14 plots the zero-state response of an RL circuit.

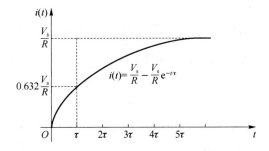

Fig. 4.14 The zero-state response of RL circuit

4.6.2 Zero-state response of RC circuit

Fig. 4.15 is a first-order RC circuit. The circuit is composed of the DC current source $i_s(t)\big|_{t\geqslant 0}=I_s$, the resistor R, the capacitor C and a switch k. The switch k has been in the position for a long time, until the switch k is closed at time $t=0$. Then find $v_C(t)$ for $t>0$.

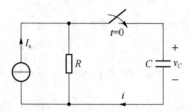

Fig. 4.15 A circuit used to illustrate the zero-state response of *RC* circuit

Before the switch k is closed, the voltage of the capacitor is zero, i. e. $v_C(0^-)=0$. When the switch k is closed, a nonhomogeneous differential equation is obtained according to the circuit KCL equation and the VCR of the capacitor

$$C\frac{dv_C(t)}{dt}+\frac{v_C(t)}{R}=i_s(t), \quad t\geqslant 0 \tag{4.37}$$

According to the initial condition is $v_C(0)=0$, the solution of the nonhomogeneous equation is given by

$$v_C(t)=RI_s(1-e^{-\frac{t}{RC}}) \tag{4.38}$$

where the term RI_s in Eq(4.38) is the steady value of the capacitor voltage $v_C(\infty)=RI_s$.

Thus, the analysis of zero-state response of a first-order circuit can be divided into three steps

(1) Calculate the stable value of the variable $v_C(\infty)$ or $i_C(\infty)$;

(2) Find the time constant τ;

(3) The variable after the switching is $v_C(t)=v_C(\infty)(1-e^{-\frac{t}{\tau}})$ or $i_C(t)=i_C(\infty)(1-e^{-\frac{t}{\tau}})$.

4.7 Complete response and three element method for *RL* and *RC* circuit

In the previous subsections, we analyze the zero-input and zero-state response of first-order circuit respectively. Commonly, the response of the circuit with both an external excitation input and the initial energy storage. In this case, the circuit response is called the complete response of the circuit. Based on the superposition theorem, the complete response is the sum of the zero-input response and the zero-state response for the linear *RC* or *RL* circuit.

As is shown in Fig. 4.16, the capacitor has initial energy storage before closing the switch k, i. e., $v_C(0^-)=V_0$, the DC voltage source is $v_s(t)\big|_{t\geqslant 0}=V_s$. Find the complete response of the circuit for $t>0$.

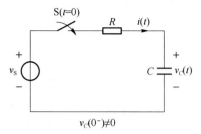

Fig. 4.16 An RC circuit

Bosed on KVL and VCR equations,

$$RC\frac{dv_C(t)}{dt}+V_C(t)=V_s \quad (4.39)$$

The particular solution is $v_{cp}(t)=V_s$ and the general solution is $v_{ch}(t)=Ae^{-\frac{t}{RC}}$.

The coefficient can be obtained according to the initial value of capacitor voltage

$$A=v_C(0^+)-V_s=V_0-V_s \quad (4.40)$$

The complete response of the capacitor voltage is

$$v_C(t)=v_{ch}(t)+v_{cp}(t)=V_s+(V_0-V_s)e^{-\frac{1}{RC}t} \quad (4.41)$$

Rewrite Eq(4.41), then

$$v_C(t)=V_0 e^{-\frac{1}{RC}t}+V_s(1-e^{-\frac{1}{RC}t}) \quad (4.42)$$

where $V_0 e^{-\frac{1}{RC}t}$ is the zero-input response of capacitor voltage and $V_s(1-e^{-\frac{1}{RC}t})$ is the zero-state response of capacitor voltage. It is verified that the complete response of a variable is equal to the sum of the zero-input response and the zero-state response.

According to the above analysis, under the excitation of DC power supply, the voltage and current of any branch can be directly solved by the following formula

$$y(t)=y(\infty)+[y(t_0^+)-y(\infty)]e^{-\frac{t-t_0}{\tau}} \quad (4.43)$$

where $y(t_0^+)$ is the initial value, $y(\infty)$ is the value in the steady state, t_0 is the switching time and τ is the time constant.

No matter for zero-input response, or zero-state response or complete response, as long as the initial value, steady-state value and time constant of the variable are known, the response results can be written directly. This method is called three element method. Table 4.1 summarizes the three element method of zero input response, zero state response and complete response of the first-order circuit.

Table 4.1 Summary of the solution of response of first-order circuit

		Solution (t_0 is the switching time)	Time constant
Zero-input response	RC circuit	$y(t) = y(t_0^+)e^{-\frac{t-t_0}{\tau}}$ $y(t)$ is $v_C(t)$ or $i_C(t)$	$\tau = RC$
	RL circuit		$\tau = \dfrac{L}{R}$
Zero-state response	RC circuit	$y(t) = y(\infty)\left(1 - e^{-\frac{t-t_0}{\tau}}\right)$ $y(t)$ is $v_C(t)$ or $i_C(t)$	$\tau = RC$
	RL circuit		$\tau = \dfrac{L}{R}$
Complete response	RC circuit	$y(t) = y(\infty) + [y(t_0^+) - y(\infty)]e^{-\frac{t-t_0}{\tau}}$	$\tau = RC$
	RL circuit		$\tau = \dfrac{L}{R}$

Q4.1 The switch in the circuit shown in Fig. 4.17 has been in position 1 for a long time. At $t=0$, the switch S moves from position 1 to position 2. Calculate the capacitor current i_C for $t>0$.

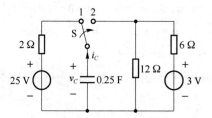

Fig. 4.17 The circuit of Q4.1

Solution: The initial voltage of the capacitor is

$$v_C(0^+) = v_C(0^-) = 25 \text{ V} \tag{4.44}$$

After the switch, we use Thevenin theorem to analyze the circuit in the right part of the capacitor, where the open circuit voltage is given by

$$v_{oc} = \frac{12 \ \Omega \times 3 \text{ V}}{12 \ \Omega + 6 \ \Omega} = 2 \text{ V} \tag{4.45}$$

and the equivalent resistor is

$$R_{eq} = 12 \ \Omega // 6 \ \Omega = 4 \ \Omega \tag{4.46}$$

The circuit can be reploted as Fig. 4.18

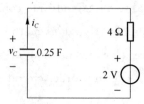

Fig. 4.18 The circuit after switching

So the steady state voltage
$$v_C(\infty) = 2 \text{ V} \tag{4.47}$$
and the time constant
$$\tau = R_{eq}C = 4 \text{ }\Omega \times 0.25 \text{ F} = 1 \text{ s} \tag{4.48}$$
Then the complete response is
$$\begin{aligned}v_C(t) &= v_C(\infty) + [v_C(0^+) - v_C(\infty)]e^{-t/\tau}\\ &= 2 + (25-2)e^{-t} = 2 + 23e^{-t} \text{ V}\end{aligned} \tag{4.49}$$
Therefore, the current across the capacitor is
$$i_C(t) = -C\frac{dv_C(t)}{dt} = -0.25 \times (-23)e^{-t} = 5.75e^{-t} \text{ A} \tag{4.50}$$

4.8 Exercises

E4.1 The switch in Fig. 4.19 has been closed for a long time before opening at $t=0$. Calculate the current i for $t>0$.

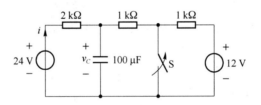

Fig. 4.19 The circuit of E4.1

E4.2 The switch in the circuit shown in Fig. 4.20 has been in position 1 for a long time. At $t=0$, the switch S moves from position 1 to position 2. Calculate the voltage $v(t)$ for $t>0$.

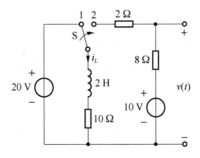

Fig. 4.20 The circuit of E4.2

E4.3 The switch in Fig. 4.21 has been open for a long time before closing at $t=0$. Calculate the voltage v_L for $t>0$.

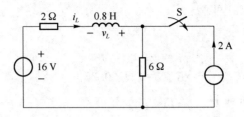

Fig. 4.21   The circuit of E4.3

E4.4   The switch in Fig. 4.22 has been open for a long time before closing at $t=0$. Calculate the current i_1 for $t>0$.

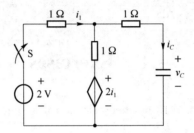

Fig. 4.22   The circuit of E4.4

E4.5   The switch in Fig. 4.23 has been open for a long time before closing at $t=0$. Calculate the voltage v_L for $t>0$.

(a) Find $i_1(0^-)$, $i_2(0^-)$, $i_1(0^+)$ and $i_2(0^+)$

(b) Calculate $i_1(t)$ and $i_2(t)$ for $t>0$.

(c) Explain why $i_2(0^-) \neq i_2(0^+)$

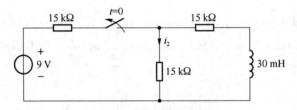

Fig. 4.23   The circuit of E4.5

E4.6   The switch in Fig. 4.24 has been closed for a long time before opening at $t=0$. Calculate the voltage $v_o(t)$ for $t>0$.

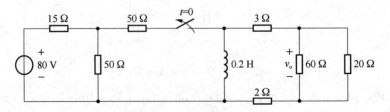

Fig. 4.24   The circuit of E4.6

E4.7 The switch in Fig. 4.25 has been open for a long time before closing at $t=0$. Calculate the voltage v_L for $t>0$.

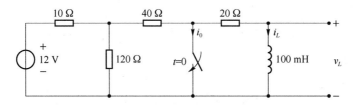

Fig. 4.25 The circuit of E4.7

E4.8 The switch in Fig. 4.26 has been in position a for a long time. At $t=0$, it moves instantaneously from a to b.

(a) find $v_o(t)$ for $t>0$.

(b) What is the total energy delivered to the 1 kΩ resistor?

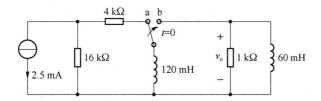

Fig. 4.26 The circuit of E4.8

E4.9 The switch in the circuit in Fig. 4.27 has been in position a for a long time. At $t=0$, the switch is thrown to position b. Calculate

(a) i, v_1 and v_2 for $t \geqslant 0$ if $v_2(0^-)=0$

(b) the energy stored in the capacitor at $t=0$

(c) the energy trapped in the circuit and the total energy dissipated in the 5 kΩ resistor if the switch remains in position b.

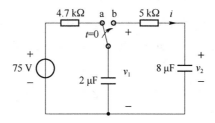

Fig. 4.27 The circuit of E4.9

E4.10 In the circuit shown in Fig. 4.28, $v_{C1}(0^-)=v_{C2}(0^-)=0$ and the switch is closed when $t=0$. Calculate $i_{C1}(0^+)$, $i_{C2}(0^+)$, $v_{C1}(\infty)$ and $v_{C2}(\infty)$.

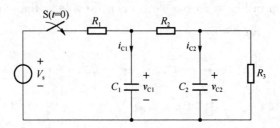

Fig. 4.28 The circuit of E4.10

E4.11 The circuit shown in Fig. 4.29 has reached steady state before the switch closes at $t=0$. Calculate the voltage $v_C(t)$ for $t>0$.

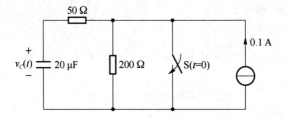

Fig. 4.29 The circuit of E4.11

E4.12 The circuit shown in Fig. 4.30 has reached steady state before the switch opens at $t=0$. Calculate the voltage $v_C(t)$ and current $i(t)$ for $t>0$.

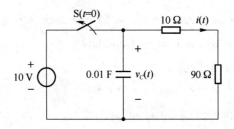

Fig. 4.30 The circuit of E4.12

E4.13 The switch in Fig. 4.31 has been position 1 for a long time. At $t=0$, the switch moves from position 1 to position 2. Calculate the inductive current i_L for $t \geqslant 0^+$.

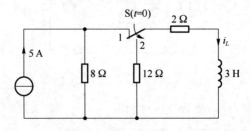

Fig. 4.31 The circuit of E4.13

Chapter 4 *RC* Circuit and *RL* Circuit

E4.14 The circuit shown in Fig. 4.32 has reached steady state at $t=0^-$. At $t=0$, switch S_1 is closed and S_2 is opened.

(a) Calculate the inductive current i_L for $t \geq 0^+$;

(b) Calculate the inductive current i_L for $t \geq 0^+$ when the voltage of the voltage source changes to 60V.

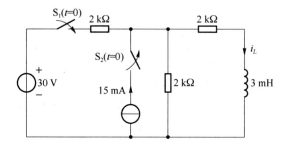

Fig. 4.32 The circuit of E4.14

E4.15 The circuit is shown in Fig. 4.33. Calculate the inductive current $i_L(t)$ for $t \geq 0^+$ under the following conditions:

(a) $v_s(t)=0$ V, $i_L(0^-)=3$ A;

(b) $v_s(t)=10$ V, $i_L(0^-)=0$;

(c) $v_s(t)=20$ V, $i_L(0^-)=-1$ A.

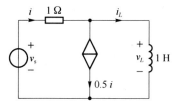

Fig. 4.33 The circuit of E4.15

E4.16 As shown in Fig. 4.34, the switch is closed at $t=0$. $v_C(0^-)=1$ V and $v_s=1$ V. Calculate the voltage $v_0(t)$ across the 2 Ω resistor.

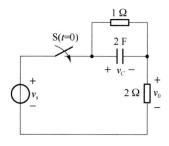

Fig. 4.34 The circuit of E4.16

E4.17 The circuit shown in Fig. 4.35 has reached steady state at $t=0$. Calculate the voltage $v_{ab}(t)$ for $t \geqslant 0$.

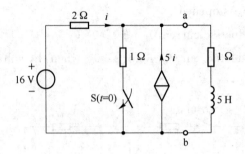

Fig. 4.35　The circuit of E4.17

E4.18　As shown in Fig. 4.36, $v_C(0^-) = 2$ V. Calculate the voltage $v_C(t)$ for $t \geqslant 0^+$.

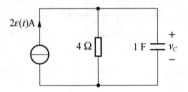

Fig. 4.36　The circuit of E4.18

E4.19　Use the unit step function to represent the three voltage waveforms shown in Fig. 4.37.

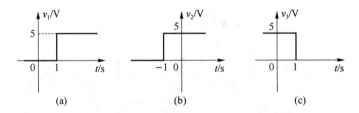

Fig. 4.37　The circuit of E4.19

E4.20　The circuit and the waveform of v_{in} are shown in Fig. 4.38. Assume that $T = 5RC$. Calculate the voltage v_o for $t > 0$.

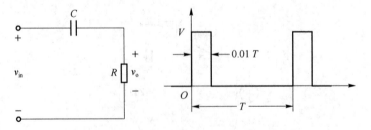

Fig. 4.38　The circuit of E4.20

Chapter 5

Sinusoidal Steady-State Analysis

So far, we have been learning the circuits with DC sources, whereas in this chapter we will consider the circuits supported by time-varying voltage or current sources (AC source). Particularly, we are interested in sources in which the value of the voltage or current varies sinusoidally. The circuit analysis methods and theorems introduced in Chapter 2 to Chapter 4 is valid for circuits with time-varying sources, so some of the results in this chapter will be familiar to you.

The goal of Chapter 5: Sinusoidal steady-state analysis

(1) understand the definition of sinusoidal signal, sinusoidal source and sinusoidal response;

(2) skill in the phase representation of sinusoidal variable;

(3) understand of the definition of impedance and admittance;

(4) understand and skill in the analysis methods and theorems in phase representation;

(5) understand complex power and its variants, and skill in their calculations.

5.1 Sinusoidal function

A sinusoidal function varies sinusoidally with time, which is determined by the amplitude, angular frequency and initial phase, as is shown in Fig. 5.1.

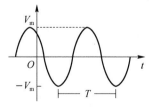

Fig. 5.1 A sinusoidal function

In this chapter, we use the cosine function as a sinusoidal function. A sinusoidal function can be written as

$$v(t) = V_m \cos(\omega t + \phi) \qquad (5.1)$$

where (1) ϕ is the initial phase angle at $t=0$. If $\phi=0$, the function is illustrated as Fig. 5.2 (a); if $\phi<0$, the function is illustrated as Fig. 5.2 (b); if $\phi>0$, the function is illustrated as Fig. 5.2 (c).

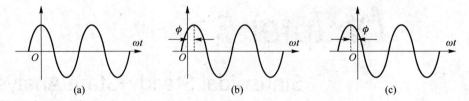

Fig. 5.2   A sinusoidal function with different initial phases

(2) ω is the angular frequency, determined by $\omega=2\pi f=\dfrac{2\pi}{T}$, where f is the frequency measured in hertz (Hz) and T is the time measured in seconds (s).

(3) V_m is the maximum amplitude.

Another important characteristic of the sinusoidal function is rms value. The rms value of a periodic function is defined as the square root of the mean value of the squared function. Hence, the rms value of $v(t)=V_m\cos(\omega t+\phi)$ is given by

$$V_{rms}=\sqrt{\frac{1}{T}\int_{t_0}^{t_0+T}v(t)^2\,dt}=\sqrt{\frac{1}{T}\int_{t_0}^{t_0+T}V_m^2\cos^2(\omega t+\phi)\,dt}=\frac{V_m}{\sqrt{2}} \qquad (5.2)$$

The rms value of the sinusoidal function depends only on the maximum amplitude V_m, regardless of frequency and phase. We stress the importance of the rms value since it relates to power calculations in the following sections.

5.2   Sinusoidal response

The circuit shown in Fig. 5.3 describe the a common circuit.

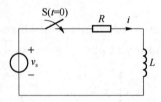

Fig. 5.3   A common circuit

The voltage provided by the voltage source is

$$v_s=V_m\cos(\omega t+\phi) \qquad (5.3)$$

The only difference with the circuits in Chapter 2 is that the voltage source here provides a time-varying sinusoidal voltage rather than a constant voltage. Applying KVL to the circuit shown in Fig. 5.3,

$$L\frac{di_L}{dt}+Ri_L=V_m\cos(\omega t+\phi) \qquad (5.4)$$

The solution to Eq(5.4) is given by

$$i_L(t) = \frac{-V_m}{\sqrt{R^2+\omega^2L^2}}\cos(\phi-\theta)e^{-(R/L)t} + \frac{V_m}{\sqrt{R^2+\omega^2L^2}}\cos(\omega t+\phi-\theta) \quad (5.5)$$

where $\theta = \arctan(\omega L/R)$. The first term on the right-hand side of Eq(5.5) is referred as the transient component of the current because it becomes infinitesimal as time elapses. The second term on the right-hand side is known as the steady-state component of the solution. The steady state response exists as long as the source continues to supply the voltage. Because the transient response vanishes as time elapses, the steady state response must satisfy the differential equation. In a linear circuit driven by sinusoidal sources, the steady state response is also a sinusoidal function. The frequency of the steady state response is identical to the frequency of the sinusoidal source. The amplitude and phase angle commonly differ from those of the sinusoidal source.

5.3 The phasor

Due to the fact that the frequencies of the variables in the same circuit are the same, a variable is represented only by the maximum value and the initial phase angle for simplification. Therefore, we convert sinusoid function into a complex number.

The phasor is a complex number that represents the amplitude and the phase angle of a sinusoidal function. The phasor concept results from Euler's formula, which transoforms the exponential function to the trigonometric function

$$e^{j\phi} = \cos\phi + j\sin\phi \quad (5.6)$$

The cosine function can be considered as the real part of the exponential function and the sine function is the imaginary part of the exponential function, that is, $\cos\phi = \text{Re}\{e^{j\phi}\}$ and $\sin\phi = \text{Im}\{e^{j\phi}\}$, where Re{·} is the real part and Im{·} is the imaginary part.

5.3.1 Phasor transformation

Based on the Euler's formula, a sinusoidal function can be expressed as

$$v(t) = V_m\cos(\omega t+\phi) = \text{Re}\{V_m e^{j(\omega t+\phi)}\} = \text{Re}\{V_m e^{j\phi} e^{j\omega t}\} \quad (5.7)$$

where $V_m e^{j\phi}$ in the last term of Eq(5.7) is a complex number consisting of the amplitude and the initial phase angle of the sinusoidal function, and $e^{j\omega t}$ represents the frequency information. Based on the definition, a complex number is the phasor representation or phasor transform of a sinusoidal function. Thus, the phasor transformation or phasor of the sinusoidal function $V_m\cos(\omega t+\phi)$ is denoted by $\dot{V}$ or $\mathscr{P}\{V_m\cos(\omega t+\phi)\}$, and

$$\mathscr{P}\{V_m\cos(\omega t+\phi)\} = V_m e^{j\phi} \quad (5.8.a)$$

or

$$\dot{V} = V_m e^{j\phi} \quad (5.8.b)$$

Therefore, the phasor transformation maps a sine function from the time domain to the complex domain (frequency domain). Phasors can also be expressed in the following forms

$$\dot{V} = V_m e^{j\phi} = V_m \angle \phi = V_m (\cos\phi + j\sin\phi) \tag{5.9}$$

Phasor transformation helps to simplify the circuit analysis under sinusoidal source, and it can be used to solve the steady-state response of a circuit with AC source.

5.3.2 Inverse phasor transformation

The phasor transformation introduced in the previous section is reversible, and its inverse transform is the inverse phasor transformation. That is, you can convert a phasor from the complex domain to the time domain. The inverse phasor transformation of the phasor $V_m e^{j\phi}$ is denoted by $\mathscr{P}^{-1}\{V_m e^{j\phi}\}$, and

$$\mathscr{P}^{-1}\{V_m e^{j\phi}\} = \mathrm{Re}\{V_m e^{j\phi} e^{j\omega t}\} = V_m \cos(\omega t + \phi) \tag{5.10}$$

where ω cannot be deduced from the phasor, but it is same to the frequency of the source.

To sum up, the phasor transformation and the inverse phasor transformation can be converted between the time domain and the complex domain.

5.3.3 Basic operations of phasors

This section introduces the basic operations of phasors, including the phasor addition (subtraction), differentiation and integration.

1. Addition and subtraction operation

If the sinusoidal function is $v_n(t) = V_{mn}\cos(\omega t + \phi_n)$, then sum of the sinusoidal functions in phasor is given by $\mathscr{P}\{\sum_{n=1}^{N} v_n(t)\} = \sum_{n=1}^{N} \mathscr{P}\{v_n(t)\}$.

Proof:

$$\sum_{n=1}^{N} v_n(t) = \sum_{n=1}^{N} V_{mn}\cos(\omega t + \phi_n)$$

$$= \sum_{n=1}^{N} \mathrm{Re}\{V_{mn} e^{j\phi_n} e^{j\omega t}\} = \mathrm{Re}\{(\sum_{n=1}^{N} V_{mn} e^{j\phi_n}) e^{j\omega t}\}$$

$$= \mathrm{Re}\{\sum_{n=1}^{N} \mathscr{P}\{v_n(t)\} e^{j\omega t}\} \tag{5.11}$$

Therefore,

$$\mathscr{P}\{\sum_{n=1}^{N} v_n(t)\} = \sum_{n=1}^{N} \mathscr{P}\{v_n(t)\} \tag{5.12}$$

Q5.1 If $y_1(t) = 20\cos(\omega t + 30°)$ and $y_2(t) = 40\cos(\omega t + 60°)$, (1) calculate $y(t) = y_1(t) + y_2(t)$ by using trigonometric identities; (2) Calculate $y(t)$ by phasor operation.

Solutions:

(1) By the trigonometric identities, we can obtain

$$y(t) = 44.72\cos(\omega t + 33.43°) \tag{5.13}$$

(2) By addition operation in phasor

$$\dot{Y} = \dot{Y}_1 + \dot{Y}_2$$
$$= 20\angle -30° + 40\angle 60°$$
$$= (17.32 - j10) + (20 + j34.64)$$
$$= 37.32 + j24.64$$
$$= 44.72\angle 33.43° \tag{5.14}$$

Based on inverse phasor transformation, we can obtain

$$y(t) = 44.72\cos(\omega t + 33.43°) \tag{5.15}$$

Subtraction operation in phasor is similar to addition, and you can prove it by yourself.

2. Differential operation

If the sinusoidal function is $v_n(t) = V_{mn}\cos(\omega t + \phi_n)$, then differential of the sinusoidal functions in phasor is given by $\mathscr{P}\left\{\dfrac{dv(t)}{dt}\right\} = j\omega \mathscr{P}\{v(t)\}$.

Proof:

$$\frac{dv(t)}{dt} = \frac{d(V_m\cos(\omega t + \phi))}{dt} = \frac{d(\mathrm{Re}\{V_m e^{j\phi} e^{j\omega t}\})}{dt}$$

$$= \mathrm{Re}\left\{\frac{d(V_m e^{j\phi} e^{j\omega t})}{dt}\right\} = \mathrm{Re}\{V_m e^{j\phi}(j\omega e^{j\omega t})\} = \mathrm{Re}\{j\omega \dot{V} e^{j\omega t}\} \tag{5.16}$$

Then,

$$\mathscr{P}\left\{\frac{dv(t)}{dt}\right\} = j\omega \mathscr{P}\{v(t)\} \tag{5.17}$$

3. Integral operation

If the sinusoidal function is $v_n(t) = V_{mn}\cos(\omega t + \phi_n)$, then integration of the sinusoidal functions in phasor is given by $\mathscr{P}\left\{\int v(t)dt\right\} = \dfrac{1}{j\omega}\mathscr{P}\{v(t)\}$.

Proof:

$$\int v(t)dt = \int V_m\cos(\omega t + \phi)dt = \int \mathrm{Re}\{V_m e^{j\phi} e^{j\omega t}\}dt$$

$$= \mathrm{Re}\left\{\int V_m e^{j\phi} e^{j\omega t} dt\right\} = \mathrm{Re}\left\{V_m e^{j\phi} \cdot \frac{1}{j\omega} e^{j\omega t}\right\} = \mathrm{Re}\left\{\frac{1}{j\omega} \dot{V} e^{j\omega t}\right\} \tag{5.18}$$

Then,

$$\mathscr{P}\left\{\int v(t)dt\right\} = \frac{1}{j\omega}\mathscr{P}\{v(t)\} \tag{5.19}$$

5.4 Passive circuit elements in phasor

It can be known from VCR of dynamic elements that the differential form cannot be written in a simple linear form like a resistor, but it can have a linear expression similar to a resistor in phasor.

The VCR of circuit elements (such as resistance, inductance, capacitance) in phasor is written as follows

$$Z = \frac{\dot{V}}{\dot{I}} \tag{5.20}$$

This equation, like Ohm's Law, expresses the linear relationship between voltage and current, so it is called Ohm's Law in phasor. The ratio of voltage phasor to current phasor in Eq(5.20) is a complex number, where Z is the impedance of the element measured in ohms.

Since the impedance may be complex, it can be written as

$$Z = R + jX \tag{5.21}$$

where R is real part and X is the imaginary part, called reactance measured in ohms.

The impedance of circuit elements (such as resistor, inductor, capacitor) is given by $Z_R = R$, $Z_L = j\omega L$, and $Z_C = \frac{1}{j\omega C} = -j\frac{1}{\omega C}$. The impedance of inductor and capacitor contains only the imaginary part. For an inductor, $X_L = \omega L$ is the reactance of the inductor; for a capacitor, $X_C = -\frac{1}{\omega C}$ is the reactance of the capacitor.

The smaller frequency ω results in smaller X_L and larger X_C. If $\omega = 0$ (i.e., the DC source), $X_L = 0$ and $X_C \to \infty$. This implies that an inductor acts like a short circuit in DC circuit; a capacitor acts like an open circuit in DC circuit.

The admittance of resistor, inductor, and capacitor is given by

$$Y = \frac{1}{Z} = \frac{\dot{I}}{\dot{V}} \tag{5.22}$$

Admittance is the reciprocal of impedance, of which unit is siemens (S). It can measure the ease with which a circuit or device allows current to flow.

Admittance, just like impedance, is a complex number, made up of a real part (the conductance G), and an imaginary part (the susceptance B), thus is given by

$$Y = G + jB \tag{5.23}$$

where G is the conductance measured in siemens (S) and B is the susceptance measured in siemens (S).

The admittance of circuit components (such as resistors, inductors, and capacitors) are given by $Y_R = G = \frac{1}{R}$, $Y_L = \frac{1}{j\omega L} = -j\frac{1}{\omega L}$, and $Y_C = j\omega C$, respectively. The admittance of the inductor and capacitor only contains the imaginary part. For an inductor, $B_L = -\frac{1}{\omega L}$ is the susceptance of the inductor; for a capacitor, $B_C = \omega C$ is the susceptance of the capacitor.

Table 5.1 The summary of the circuit elements in phasor

Element	Impedance	Reactance	Admittance	Susceptance
Resistor	R	—	G	—
Inductor	$j\omega L$	ωL	$j(-1/\omega L)$	$-1/\omega L$
Capacitor	$j(-1/\omega C)$	$-1/\omega C$	$j\omega C$	ωC

5.5 VCR of circuit element in phasor

5.5.1 VCR of resistor in phasor

According to Ohm's law, if the current of a resistor varies sinusoidally with the time, denoted by $i(t)=I_m\cos(\omega t+\phi)$, the voltage of the resistor is

$$v(t)=RI_m\cos(\omega t+\phi) \tag{5.24}$$

The voltage in phasor is

$$\dot{V}=RI_m e^{j\phi}=RI_m\angle\phi \tag{5.25}$$

Therefore, the VCR of a resistor in phasor is written as

$$\dot{V}=R\cdot\dot{I} \tag{5.26}$$

which states that the voltage in phasor at the terminals of a resistor is the product of the resistance and the current in phasor. Fig. 5.4 shows the VCR of a resistor. We can find that the voltage and the current of a resistor are in the same phase.

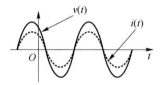

Fig. 5.4 The relationship between the current and voltage at the terminal of a resistor

5.5.2 VCR of capacitor in phasor

If the voltage at the terminal of a capacitor is $v(t)=V_m\cos(\omega t+\phi)$, the VCR of a capacitor is

$$i(t)=C\frac{dv}{dt}=\omega CV_m\cos(\omega t+\phi+90°) \tag{5.27}$$

Therefore, the VCR of a capacitor in phasor is

$$\dot{V}=\frac{1}{j\omega C}\dot{I} \tag{5.28}$$

which states that the voltage across the capacitor in phasor is the product of the capacitor's impedance and the current in phasor.

Fig. 5.5 shows the relationship between the current and the voltage of a capacitor. We can find that the current leads the voltage by 90°.

5.5.3 VCR of inductor in phasor

If the current flowing in an inductor is $i(t)=I_m\cos(\omega t+\phi)$, the VCR of an inductor is

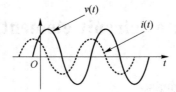

Fig. 5.5 The relationship between the current and voltage at the terminal of a capacitor

$$v(t) = L\frac{di}{dt} = \omega L I_m \cos(\omega t + \phi + 90°) \tag{5.29}$$

There, the VCR of an inductor in phasor is

$$\dot{V} = j\omega L \dot{I} \tag{5.30}$$

which states that the voltage of the inductor in phasor is the product of the inductor's impedance and the current in phasor.

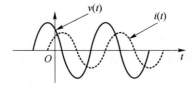

Fig. 5.6 The relationship between the current and voltage at the terminal of an inductor

Fig. 5.6 shows the relationship between the current and the voltage of an inductor. We can find that the voltage leads the current by 90°.

5.6　Kirchhoff's Law in phasor

According to Kirchhoff's law of current (KCL) and Kirchhoff's law of voltage (KVL) studied in Chapter 2, their phasor forms will be rewritten in this section.

5.6.1　KCL in phasor

According to KCL in Chapter 2, the KCL in phasor is given by

$$\sum_{n=1}^{N} \dot{I}_n = 0 \tag{5.31}$$

Proof:

$$\sum_{n=1}^{N} i_n(t) = \sum_{n=1}^{N} \mathrm{Re}[I_{mn} e^{j\theta_n} e^{j\omega t}] = \mathrm{Re}\left[e^{j\omega t} \cdot \sum_{n=1}^{N} I_{mn} e^{j\theta_n}\right] = \mathrm{Re}\left[e^{j\omega t} \cdot \sum_{n=1}^{N} \dot{I}_n\right] = 0 \tag{5.32}$$

Due to $e^{j\omega t} \neq 0$, so $\sum_{n=1}^{N} \dot{I}_n = 0$. KCL in phasor holds that the algebraic sum of all the currents in phasor in any node in a circuit is zero.

5.6.2 KVL in phasor

According to KVL in Chapter 2, the KVL in phasor is given by

$$\sum_{n=1}^{N} \dot{V}_n = 0 \tag{5.33}$$

Proof:

$$\sum_{n=1}^{N} v_n(t) = \sum_{n=1}^{N} \text{Re}[V_{mn} e^{j\theta_n} e^{j\omega t}] = \text{Re}\Big[\sum_{n=1}^{N} V_{mn} e^{j\theta_n} e^{j\omega t}\Big] = \text{Re}\Big[e^{j\omega t} \cdot \sum_{n=1}^{N} \dot{V}_n\Big] = 0 \tag{5.34}$$

Due to $e^{j\omega t} \neq 0$, so $\sum_{n=1}^{N} \dot{V}_n = 0$. KVL in phasor holds that the algebraic sum of all the phasor voltages around any loop in a circuit is zero.

5.7 Circuit simplifications in phasor

5.7.1 Impedances in series and parallel

Multiple impedances in series can be equivalent to an impedance by simply adding the individual impedance. The only difference is that combing impedances involves the algebraic operation of complex numbers.

As Fig. 5.7 shows, the impedance Z_1, Z_2, ⋯ and Z_N are connected in series between terminals a and b. That means they carry the same current in phasor $\dot{I}$. By KVL in phasor,

$$\dot{V}_{ab} = \dot{V}_1 + \dot{V}_2 + \cdots + \dot{V}_N = \sum_{n=1}^{N} \dot{V}_n = \sum_{n=1}^{N} (Z_n \dot{I}) = \dot{I} \sum_{n=1}^{N} Z_n \tag{5.35}$$

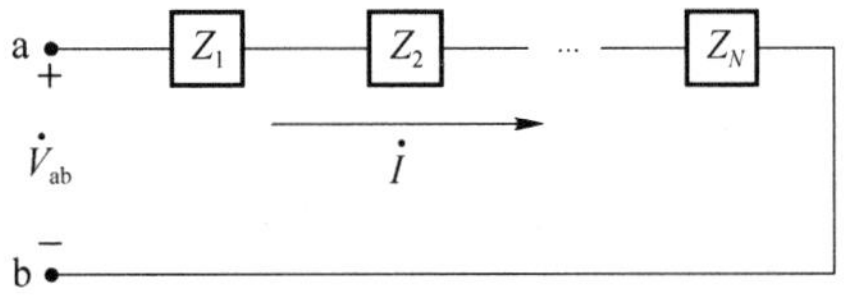

Fig. 5.7 Impedances in series

The equivalent impedance between terminals a and b is

$$Z_{ab} = \frac{\dot{V}_{ab}}{\dot{I}_{ab}} = \sum_{n=1}^{N} Z_n \tag{5.36}$$

Impedance connected in parallel may be reduced to a single equivalent impedance by the reciprocal relationship

As Fig. 5.8 shows, the impedance Z_1, Z_2, ⋯, and Z_N are in parallel between terminals a and b. That means they have the same voltage in phasor $\dot{V}_{ab}$ across their terminals. By KCL in phasor,

$$i = i_1 + i_2 + \cdots + i_N = \sum_{n=1}^{N} \dot{i}_n = \sum_{n=1}^{N} \left(\frac{\dot{V}_{ab}}{Z_n}\right) = \dot{V}_{ab} \sum_{n=1}^{N} \left(\frac{1}{Z_n}\right) \quad (5.37)$$

Fig. 5.8 Impedances in parallel

The equivalent impedance between terminals a and b can be expressed as

$$\frac{1}{Z_{ab}} = \frac{\dot{i}}{\dot{V}_{ab}} = \sum_{n=1}^{N} \frac{1}{Z_n} \quad (5.38)$$

which can be rewritten as the admittance form

$$Y_{ab} = \sum_{n=1}^{N} Y_n \quad (5.39)$$

5.7.2 Source transformations in phasor, Thevenin theorem in phasor and Norton theorem

The source transformation, Thevenin theorem and Norton theorem introduced in Chapter 3 can also be applied in the complex domain and be written in phasor expressions. Fig. 5.9 shows Thevenin theorem and Norton theorem in phasor using source transformation, which is given by

$$\begin{cases} Z_N = Z_{TH} \\ \dot{V}_{TH} = Z_{TH} \dot{I}_N \end{cases} \quad (5.40)$$

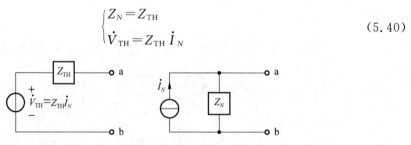

Fig. 5.9 A source transformation in the frequency domain

Thevenin's theorem in phasor states that a linear network with terminal a and b can be equivalent to an impedance in series with an independent voltage source, where the voltage value of the voltage source $\dot{V}_{TH}$ is equal to the voltage when terminal a and b are open, and the impedance Z_{TH} is equal to the equivalent impedance of the network with terminal a and b when all independent power sources in the network are set to zero.

Norton's theorem in phasor states that a linear network with terminal a and b can be equivalent to a current source in parallel with an impedance, where the current value of the current source $\dot{I}_N$ is equal to the current when terminal a and b are short-circuited, and the

Chapter 5 Sinusoidal Steady-State Analysis

impedance Z_N is equal to the equivalent impedance of the network with terminal a and b when all independent power sources in the network are set to zero.

The conversion relationship in phasor between Thevenin's Theorem and Norton's Theorem is shown in Eq(5.40).

5.8 Basic circuit analysis methods in phasor

5.8.1 Node voltage method in phasor

In Chapter 2, the node voltage method has been introduced. The principle of this method can also be used for the analysis in phasor. Let's take an example to explain the node voltage method in phasor.

Q5.2 Find the current of $\dot{I}_a$, $\dot{I}_b$ and $\dot{I}_c$ for the circuit shown in Fig. 5.10 by node voltage method.

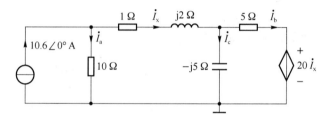

Fig. 5.10 The circuit of Q5.2

Solution:

Using KCL and Ohm's law, we can obtain

$$\begin{cases} \dfrac{\dot{V}_1}{10} + \dfrac{\dot{V}_1 - \dot{V}_2}{1+j2} - 10.6 = 0 \\ \dfrac{\dot{V}_2 - \dot{V}_1}{1+j2} + \dfrac{\dot{V}_2}{-j5} + \dfrac{\dot{V}_2 - 20\dot{I}_x}{5} = 0 \\ \dot{I}_x = \dfrac{\dot{V}_1 - \dot{V}_2}{1+j2} \end{cases} \quad (5.41)$$

So, the node voltages are given by

$$\begin{cases} \dot{V}_1 = 68.4 - j16.8 \text{ V} \\ \dot{V}_2 = 68 - j26 \text{ V} \end{cases} \quad (5.42)$$

Hence the required currents are

$$\begin{cases} \dot{I}_a = \dfrac{\dot{V}_1}{10} = 6.84 - j1.68 \text{ A} \\[6pt] \dot{I}_x = \dfrac{\dot{V}_1 - \dot{V}_2}{1+j2} = 3.76 + j1.68 \text{ A} \\[6pt] \dot{I}_b = \dfrac{\dot{V}_2 - 20\dot{I}_x}{5} = -1.44 - j11.92 \text{ A} \\[6pt] \dot{I}_c = \dfrac{\dot{V}_2}{-j5} = 5.2 + j13.6 \text{ A} \end{cases} \qquad (5.43)$$

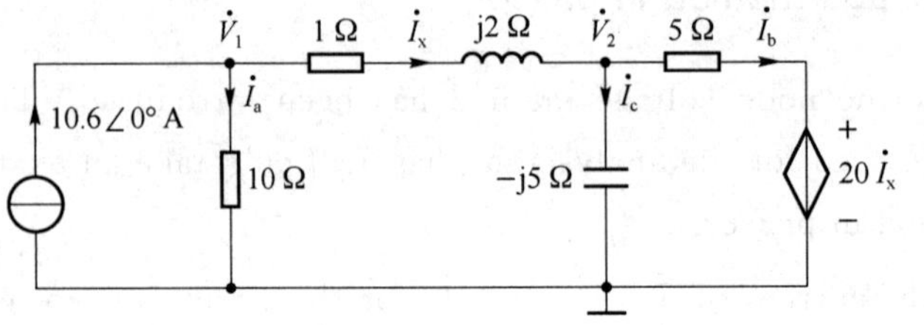

Fig. 5.11 The circuit of Q5.2 with node voltages

5.8.2 Mesh current method in phasor

In Chapter 2, the mesh current method has been introduced. The principle of this method can also be used for the analysis in phasor. Let's take an example to explain the mesh current method in phasor.

Q5.3 Find the current $i_1(t)$ and $i_2(t)$ in the circuit shown in Fig. 5.12 by mesh current method.

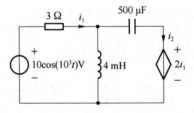

Fig. 5.12 The circuit of Q5.3

Solution:

We write the variable in phasor, $\omega = 10^3$ rad/s, $\dot{V}_s = 10\angle 0°$, $Z_R = R = 3\ \Omega$, $Z_L = j\omega L = j10^3 \times 4 \times 10^{-3} = j4\ \Omega$, and $Z_C = \dfrac{1}{j\omega C} = \dfrac{1}{j10^3 \times 500 \times 10^{-6}} = -j2\ \Omega$.

Using KVL and Ohm's law, we can obtain

$$\begin{cases} 3\dot{I}_1 + j4(\dot{I}_1 - \dot{I}_2) = 10\angle 0° \\ j4(\dot{I}_2 - \dot{I}_1) - j2\dot{I}_2 = -2\dot{I}_1 \end{cases} \qquad (5.44)$$

So, the mesh currents are given by

Chapter 5   Sinusoidal Steady-State Analysis

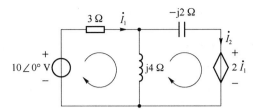

Fig. 5.13   The circuit of Q5.3 with mesh currents

$$\begin{cases} \dot{I}_1 = \dfrac{10}{7-j4} = 1.24\angle 29.7° \text{ A} \\ \dot{I}_2 = \dfrac{20+j30}{13} = 2.77\angle 56.3° \text{ A} \end{cases} \tag{5.45}$$

Hence the currents in time domain are

$$\begin{cases} i_1 = 1.24\cos(10^3 t + 29.7°) \text{ A} \\ i_2 = 2.77\cos(10^3 t + 56.3°) \text{ A} \end{cases} \tag{5.46}$$

5.9   Power calculation

In the first chapter, we have learned the definition of power, and introduced the power calculation method of DC source. Because AC circuit is more complicated than DC circuit, the power calculation in AC circuit is not as simple as that in DC circuit. This section uses a sinusoidal steady-state circuit as an example to introduce several power-related concepts, including instantaneous power, average power, reactive power, complex power, and apparent power.

At the beginning of this chapter, the root mean square of the sinusoidal function has been introduced, that is, if $v(t) = V_m \cos(\omega t + \phi)$, the root mean square of $v(t)$ is

$$V_{\text{rms}} = \sqrt{\dfrac{1}{T} \int_{t_0}^{t_0+T} V_m^2 \cos^2(\omega t + \phi) \, dt} = \dfrac{V_m}{\sqrt{2}} \tag{5.47}$$

where T is the period of the sinusoidal function. We stress the importance of the rms value as it relates to the following power calculations.

When the voltage or current source is not DC source, the power is also time-varying. To represent this characteristic, power is more delicately defined as instantaneous power.

Following the passive sign convention, the instantaneous power of an element or a single-port network is defined as

$$\begin{aligned} p(t) &= v(t)i(t) = V_m \cos(\omega t + \theta_v - \theta_i) \cdot I_m \cos(\omega t) \\ &= \dfrac{V_m I_m}{2}\cos(\theta_v - \theta_i) + \dfrac{V_m I_m}{2}\cos(2\omega t + \theta_v - \theta_i) \\ &= \dfrac{V_m I_m}{2}\cos(\theta_v - \theta_i) + \dfrac{V_m I_m}{2}\cos(\theta_v - \theta_i)\cos(2\omega t) - \dfrac{V_m I_m}{2}\sin(\theta_v - \theta_i)\sin(2\omega t) \\ &= P + P\cos(2\omega t) - Q\sin(2\omega t) \end{aligned} \tag{5.48}$$

where $P = \dfrac{V_m I_m}{2}\cos(\theta_v - \theta_i)$ and $Q = \dfrac{V_m I_m}{2}\sin(\theta_v - \theta_i)$.

Instantaneous power represents the power at a certain moment, which is of little practical significance and is not easy to measure, so the concept of average power is introduced. Average power is the average value of instantaneous power over a period of time, given by

$$P = \frac{1}{T}\int_{t_0}^{t_0+T} p(t)\mathrm{d}t = \frac{V_m I_m}{2}\cos(\theta_v - \theta_i) \qquad (5.49)$$

The average power P in Eq(5.49) is the same as that in Eq(5.48). Average power, also called active power, is the power actually consumed or produced by a circuit element or network.

Rewrite Eq(5.48),

$$p(t) = P[1 + \cos(2\omega t)] - Q\sin(2\omega t) \qquad (5.50)$$

we found that Eq(5.50) is always no less than zero, which is the irreversible part of the instantaneous power and is considered as the consumed power; the second term is a sinusoidal function with alternating sign, which is the reversible part of the instantaneous power and represents the energy exchange between circuit element (or network) and external power source. An concept to represent energy exchange is defined as reactive power,

$$Q = \frac{V_m I_m}{2}\sin(\theta_v - \theta_i)$$

In order to distinguish it from active power, the unit of reactive power is defined as Var.

The power is analyzed in the time domain above. When the voltage and current are in phasor, the voltage is given by $\dot{V} = V\angle\theta_v$, the current is given by $\dot{I} = I\angle\theta_i$, and their multiplication is given by

$$\dot{V}\dot{I} = V\angle\theta_v \cdot I\angle\theta_i = VI\angle(\theta_v + \theta_i) = VI\cos(\theta_v + \theta_i) + jVI\sin(\theta_v + \theta_i) \qquad (5.51)$$

Comparing (5.51) with (5.48), it can be seen that the difference lies in the addition and subtraction of phases. In the complex field, this problem can be solved by conjugate. The complex power is defined as

$$\widetilde{S} = \dot{V}\dot{I}^* = V\angle\theta_v \cdot I\angle-\theta_i = VI\angle(\theta_v - \theta_i) = VI\cos(\theta_v - \theta_i) + jVI\sin(\theta_v - \theta_i) = P + jQ$$
$$(5.52)$$

Eq(5.52) shows that the real part of the complex power is the average power (or active power) P, and the imaginary part is the reactive power Q.

The modulus of a complex power is the apparent power, given by

$$|\widetilde{S}| = VI = \frac{V_m I_m}{2} = \sqrt{P^2 + Q^2} \qquad (5.53)$$

where V and I are the rms of voltage and current, respectively. The unit of apparent power is the same as the complex power, which is volt-ampere (VA).

Table 5.2 summarizes the different kinds of powers and their units.

Chapter 5 Sinusoidal Steady-State Analysis

Table 5.2　Different kinds of powers and their units

Quantity	Units
Instantaneous power	Volt-amps (VA)
Complex power	Volt-amps (VA)
Apparent power	Volt-amps (VA)
Average power	Watts (W)
Reactive power	Vars

5.10　Maximum power transfer theorem in phasor

In Chapter 3, the maximum power transfer theorem has been introduced. The principle of this theorem can also be used for the analysis in phasor. Let's take an example to explain the maximum power transfer theorem in phasor.

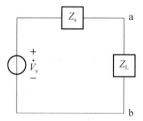

Fig. 5.14　Maximum power transfer theorem in phasor

We assume $Z_s = R_s + jX_s$ and $Z_L = R_L + jX_L$, then the complex power of the load is

$$\tilde{S} = \dot{V}\dot{I}^* = Z_L \dot{I}\dot{I}^* = |\dot{I}|^2 (R_L + jX_L) \tag{5.54}$$

The average power of the load

$$P_L = |\dot{I}|^2 R_L = \left|\frac{\dot{V}_s}{Z_L + Z_s}\right|^2 R_L = \left|\frac{\dot{V}_s}{(R_L + R_s) + j(X_L + X_s)}\right|^2 R_L$$

$$= \frac{|\dot{V}_s|^2 R_L}{(R_L + R_s)^2 + (X_L + X_s)^2} \tag{5.55}$$

In order to maximize the power delivered to the load Z_L, let

$$\begin{cases} \dfrac{\partial P_L}{\partial X_L} = 0 \\ \dfrac{\partial P_L}{\partial R_L} = 0 \end{cases} \tag{5.56}$$

then we obtain $\begin{cases} X_L = -X_s \\ R_L = R_s \end{cases}$, i.e., $Z_L = Z_s^*$.

Thus, when the load's impedance is the conjugate of the internal impedance of the source (i.e., $Z_L = Z_s^*$), the load can obtain the maximum power,

$$P_{L,\max} = \frac{|\dot{V}_s|^2}{4R_s} \tag{5.57}$$

also called conjugate matching.

5.11 Exercises

E5.1 The voltage across the terminal of the 5 μF capacitor is $30\cos(4\,000t+25°)$ V. Calculate (1) the capacitive reactance;

(2) the impedance of the capacitor;

(3) the phasor current $\dot{I}$;

(4) the steady-state expression for $i(t)$.

E5.2 Calculate the average power and the reactive power at the terminals of the network if $v = 20\cos(\omega t + 15°)$ and $i = 4\sin(\omega t - 15°)$. State whether the network is absorbing or delivering average power.

E5.3 Consider the sinusoidal voltage $v = 100\sin(628t - 30°)$ V

(1) what is the maximum amplitude of the voltage?

(2) what is the effective value of the voltage?

(3) what is the angular frequency of the voltage?

(4) what is the period in seconds?

(5) what is the phase angle in radians?

E5.4 Consider the sinusoidal voltage $v = 180\cos(20\pi t - 60°)$ V

(1) what is the maximum amplitude of the voltage?

(2) what is the frequency in hertz (Hz)?

(3) what is the frequency in radians per second (rad/s)?

(4) what is the phase angle in radians?

(5) what is the phase angle in degrees?

(6) what is the period in seconds?

E5.5 Given two sinusoidal voltages: $v_1 = 60\sin(\omega t - 30°)$ V and $v_2 = 10\cos\omega t$ V. Try to determine the phase relationship between the two voltages.

E5.6 Assume that $\dot{V} = 12\angle 0°$ V, $\dot{I} = 5\angle -36.9°$ A and $R = 3\ \Omega$ in the sinusoidal AC circuit shown in Fig. 5.15. Calculate $\dot{I}_L$ and ωL.

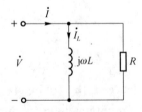

Fig. 5.15 The circuit of E5.6

E5.7 Assume that the effective values of $\dot{I}$ and $\dot{I}_2$ are 10 A and 6 A respectively, in the sinusoidal AC circuit shown in Fig. 5.16. Calculate $\dot{I}_1$.

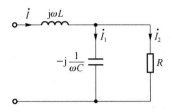

Fig. 5.16 The circuit of E5.7

E5.8 Assume that the effective values of $\dot{I}_1$ and $\dot{I}_2$ are 4 A and 3 A respectively, in the sinusoidal AC circuit shown in Fig. 5.17. Calculate $\dot{I}$.

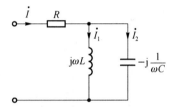

Fig. 5.17 The circuit of E5.8

E5.9 Assume that, in the sinusoidal AC circuit shown in Fig. 5.18, the effective values of each current are $I=5$ A, $I_R=5$ A and $I_L=3$ A, respectively. Calculate $\dot{I}_C$.

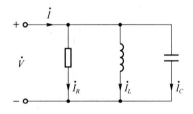

Fig. 5.18 The circuit of E5.9

E5.10 Assume that $\dot{I}_R = 2\angle -\dfrac{\pi}{3}$ A in the circuit shown in Fig. 5.19, which is a part of a sinusoidal AC circuit. Calculate $\dot{I}_L$.

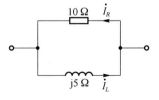

Fig. 5.19 The circuit of E5.10

E5.11 Calculate $\dot{I}$ in the sinusoidal AC circuit shown in Fig. 5.20.

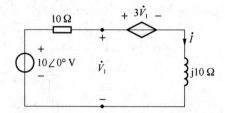

Fig. 5.20　The circuit of E5.11

E5.12 The phasor current $\dot{I}_b$ in the circuit shown in Fig. 5.21 is $5\angle 45°$.

(a) find $\dot{I}_a$, $\dot{I}_c$ and $\dot{V}_g$;

(b) if $\omega=800$ rad/s, write the expressions for $i_a(t)$, $i_c(t)$ and $v_g(t)$.

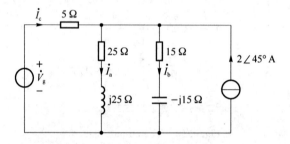

Fig. 5.21　The circuit of E5.12

E5.13 Calculate the admittance Y of the circuit shown in Fig. 5.22.

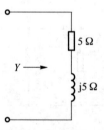

Fig. 5.22　The circuit of E5.13

E5.14 Assume that $R=1\ \Omega$, $L=1$ H and $\omega=1$ rad/s in the circuit shown in Fig. 5.23. Calculate the impedance Z of this circuit.

Fig. 5.23　The circuit of E5.14

E5.15 While $\omega=0$ and $\omega=\infty$, calculate the impedance Z_{ab} of the circuit shown in Fig. 5.24, respectively.

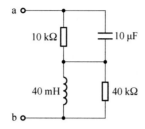

Fig. 5.24 The circuit of E5.15

E5.16 The sinusoidal voltage source in the circuit in Fig. 5.25 is developing a voltage equal to $247.49\cos(1\,000t+45°)$ V

(1) Find the Thevenin voltage with respect to the terminals a and b;

(2) Find the Thevenin impedance with respect to the terminals a and b;

(3) Draw the Thevenin equivalent.

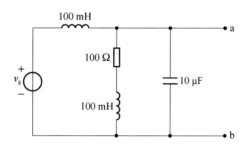

Fig. 5.25 The circuit of E5.16

E5.17 Assume that $v(t)=30\sqrt{2}\sin(\omega t-30°)$ V and $\omega=10^3$ rad/s in the sinusoidal AC circuit shown in Fig. 5.26. Calculate $i_1(t)$, $i_2(t)$, $i_3(t)$ and $i(t)$, respectively.

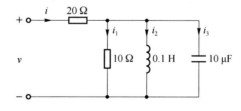

Fig. 5.26 The circuit of E5.17

E5.18 Find the average power, the reactive power, and the apparent power absorbed by the load in the circuit in Fig. 5.27 if i_g equals $30\cos100t$ mA.

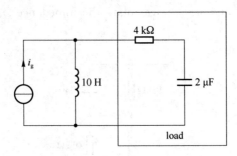

Fig. 5.27 The circuit of E5.18

E5.19 Calculate the average power supplied by the source in the sinusoidal AC circuit shown in Fig. 5.28.

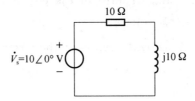

Fig. 5.28 The circuit of E5.19

E5.20 Assume that $R = \omega L = \dfrac{1}{\omega C} = 100\ \Omega$ and $\dot{I}_R = 2\angle 0°$ A in the sinusoidal AC circuit shown in Fig. 5.29. Calculate $\dot{V}_s$ and the active power supplied by the source, respectively.

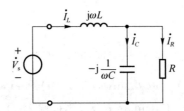

Fig. 5.29 The circuit of E5.20

Chapter 6

Experiment

In this chapter, software or hardware experiments are conducted to verify the circuit analysis methods and circuit laws we have learned in the previous chapters.

The goal of Chapter 6:

(1) learn to use basic circuit measuring instruments;

(2) learn to use software for circuit simulation;

(3) verify circuit analysis methods and circuit theorems.

6.1 Basic electronic measuring devices

6.1.1 Multi-meter

A multimeter or a multitester, also known as a VOM (volt-ohm-milliammeter), is an electronic measuring instrument that combines several measurement functions in one unit. A typical multimeter can measure voltage, current, and resistance. A multimeter can be a hand-held device useful for basic fault finding and field service work, or a bench instrument which can measure to a very high degree of accuracy.

The multimeter is used as follows:

(1) Before using the multimeter, an important thing is to perform "mechanical zero adjustment". If meter pointer does not indicate 0, adjust mechanical zero for 0 indication;

(2) During using the multimeter, do not touch the metal part of the test lead with your hand. This can ensure the accuracy of the measurement on the one hand and the safety of the person on the other hand;

(3) When measuring an element, you cannot shift gears during the measurement, especially when measuring high voltage or high current, otherwise the multimeter will be damaged. If you need to shift gears, you should disconnect the test lead first and then shifting gears;

(4) When the multimeter is in use, it must be placed horizontally to avoid causing errors. At the same time, avoid the impact of external magnetic fields on the multimeter.

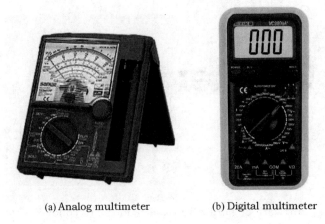

(a) Analog multimeter (b) Digital multimeter

Fig. 6.1　Multimeter

6.1.2　DC power supply

A power supply is an electrical device that supplies electric power to an electrical load. The primary function of a power supply is to convert electric current from a source to the correct voltage, current, and frequency to power the load. As a result, power supplies are sometimes referred to as electric power converters. Other functions that power supplies may perform include limiting the current drawn by the load to safe levels, shutting off the current in the event of an electrical fault, power conditioning to prevent electronic noise or voltage surges on the input from reaching the load, power-factor correction, and storing energy so it can continue to power the load in the event of a temporary interruption in the source power (uninterruptible power supply).

A DC power supply is one that supplies a constant DC voltage to its load. Depending on its design, a DC power supply may be powered from a DC source or from an AC source.

6.1.3　Resistor and slide rheostat

A resistor is a passive two-terminal electrical component that implements electrical resistance as a circuit element. In electronic circuits, resistors are used to reduce current flow, adjust signal levels, divide voltages, bias active elements, and terminate transmission lines, etc. Fixed-value resistors have resistances that only change slightly with temperature, time or operating voltage. Variable-value resistors, commonly called slide rheostat, can be used to adjust circuit elements (such as a volume control or a lamp dimmer), or as sensing devices for heat, light, humidity, force, or chemical activity. Resistors are also implemented within integrated circuits.

The electrical function of a resistor is specified by its resistance: common commercial resistors are manufactured over a range of more than nine orders of magnitude. The nominal value of the resistance falls within the manufacturing tolerance, indicated on the component.

The color ring resistance is a common discrete component in circuits. The color ring is used to represent color and error. The basic units of this type of resistance are: Ohm (Ω), kiloohm (kΩ), megohm (MΩ). The specified colors are black, brown, red, orange, yellow, green, blue, purple, gray, white, gold, silver, and colorless. The table 6.1 shows the values of each color.

Table 6.1 The resistor color code

	Silver	Gold	Black	Brown	Red	Orange	Yellow	Green	Blue	Purple	Grey	White	Colorless
Value	—	—	0	1	2	3	4	5	6	7	8	9	—
Magnitude	10^{-2}	10^{-1}	10^{0}	10^{1}	10^{2}	10^{3}	10^{4}	10^{5}	10^{6}	10^{7}	10^{8}	10^{9}	—
Allowable deviation(%)	±10	±5	—	±1	±2	—	—	±0.5	±0.25	±0.1	±0.05	—	±20

6.1.4 Capacitor and inductor

A capacitor is a passive two-terminal electronic component that stores electrical energy in an electric field. The effect of a capacitor is known as capacitance. While some capacitance exists between any two electrical conductors in proximity in a circuit, a capacitor is a component designed to add capacitance to a circuit. It is likely for capacitors to explode if the two polarity connected oppositely or the voltage is ultra-high.

An inductor is a passive two-terminal electrical component that stores energy in a magnetic field when electric current flows through it. An inductor typically consists of an insulated wire wound into a coil around a core.

6.1.5 Oscilloscope

An oscilloscope, previously called a recorder, is a type of electronic test instrument that graphically displays varying signal voltages, usually as a two-dimensional plot of one or more signals as a function of time. Other signals (such as sound or vibration) can be converted to voltages and displayed. Oscilloscopes display the change of an electrical signal over time, with voltage and time as the Y- and X-axes, respectively, on a calibrated scale.

6.1.6 Digital signal generator

A digital signal generator is an electronic device that generates repeating or non-repeating electronic signals in the digital domain. It is generally used in designing, testing, trouble shooting, and repairing electronic or electroacoustic devices, though it often has artistic uses as well.

There are many different types of signal generators with different purposes and applications and at varying levels of expense. These types include function generators, RF and microwave signal generators, pitch generators, arbitrary waveform generators, digital pattern generators and frequency generators. In general, no device is suitable for all possible applications. The commonly used one is the function generator. A function generator is a device which produces simple repetitive waveforms.

Such devices contain an electronic oscillator, a circuit that is capable of creating a repetitive waveform. Modern devices may use digital signal processing to synthesize waveforms, followed by a digital to analog converter, or DAC, to produce an analog output. The most common waveform is a sine wave, but step (pulse), square, and triangular waveform oscillators are commonly available as are arbitrary waveform generators (AWGs).

6.2 The introduction of circuit simulation software

Circuit simulators can be used for circuit drawing, circuit design and analysis as well. Following introduces some common circuit simulation software.

MultiSim is a circuit simulation software from National instruments. Although student versions always comes with limited access, it is a great simulation tool for beginners in electronics. It enables you to capture circuits, create layouts, analyse circuits and simulation. Highlight features include exploring breadboard in 3D before lab assignment submission, create printed circuit boards (PCB), etc. Breadboard simulation is possible with Multisim circuit simulator.

NgSpice-one of the popular and widely used free softwares, provides open-source circuit simulator. Ngspice is a part of g EDA project which is growing every day with the suggestions from its users. As it is a collaborative project, you can suggest improvement of the circuit simulator and be a part of the development team.

In addition, there are some other softwares that you can learn by yourself.

6.3 Experiment instances

6.3.1 DC characteristics of resistors

1. The objectives of this experiment
- Learn to use multimeters and DC voltage supplies;
- Construct, measure, and analyze simple passive element circuits.

2. The experiment equipment

In this experiment, you will learn how to connect simple circuits, how to build a circuit based upon a circuit schematic diagram, and how to perform simple measurements of each element's resistance using a multimeter. The experiment equipment includes DC voltage supplies, multimeter, resistor, slide rheostat and wires.

3. The methods, principles or laws involved in this experiment

- Ohm's law states that in a circuit, the current passing through a resistor is directly proportional to the voltage at both ends of the resistor. Introducing the constant of proportionality, the resistance is depicted by the usual mathematical equation below.

$$I = \frac{V}{R} \tag{6.1}$$

where I is the current through the resistor in units of amperes (A), V is the voltage measured across the resistor in units of volts (V), and R is the resistance of the resistor in units of ohms (Ω). In addition, ohm's law states that the resistance of the resistor R in Eq(6.1) is a constant, which is independent of the current passing through it.

- Point-by-point test method is to use a parameter as a reference, by changing the value of other parameters to measure the voltage or current accordingly.

- Resistance

Colour	code
Brown	1
Red	2
Orange	3
Yellow	4
Green	5
Blue	6
Violet	7
Grey	8
White	9
Black	0

Colour	Tolerance
Gold	5%
Silver	10%
Coloruless	20%

4. Precautions

When replacing the components, please turn off the DC power supply first.

5. Resistance from I-V data

Measure the volt-ampere characteristic of resistor $R = 200\ \Omega$.

(1) Connect the circuit according to the circuit shown in Fig. 6.2;

(2) Change the value of the DC voltage supply; use the multimeter to measure the corresponding current I and the voltage at each end of the resistance V; record the results in Table 6.2 and calculate the theoretical value of R_{th};

(3) Use the multimeter to measure R, and compare the measured result of R with the theoretical value R_{th}.

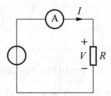

Fig. 6.2　A simple circuit

Table 6.2　Measurement for the circuit of Fig. 6.2

No. of Experiments	V/V	I/A	R_{th}/Ω
1			
2			
3			

R=_____

6. Characteristics of resistors in series/parallel

(1) Series circuit

Measure the volt-ampere characteristic of the $R_1 = 200\ \Omega$ and sliding rheostat R_2. The circuit in Fig. 6.3 shows the connection of several resistors.

Change the value of the sliding rheometer slider; use the multimeter to measure the corresponding value R_1, R_2, V_1, V_2, V, I and record them in Table 6.3.

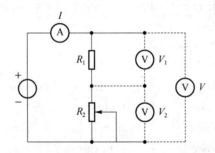

Fig. 6.3　A simple series circuit

Table 6.3　Measurement for the circuit of Fig. 6.3

No. of Experiments	V_1/V	V_2/V	V/V	R_1/Ω	R_2/Ω	I/A
1						
2						
3						

(2) Parallel circuit

Measure the volt-ampere characteristic of $R_0 = 200\ \Omega$, $R_1 = 200\ \Omega$ and sliding rheostat R_2. The circuit in Fig. 6.4 shows the connection of several resistors.

Change the value of the sliding rheometer slider; use the multimeter to measure the corresponding R_2, V_0, V_p, V, I_1, I_2, I and record them in Table 6.4.

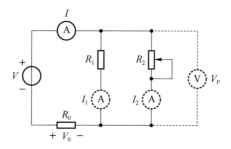

Fig. 6.4 A simple parallel circuit

Table 6.4 Measurement for the circuit of Fig. 6.4

No. of Experiments	V_0/V	V_p/V	V/V	R_2/Ω	I/A	I_1/A	I_2/A
1							
2							
3							

6.3.2 Verification for Kirchhoff's laws

1. The objective of this experiment

- Verify Kirchhoff's Laws and superposition theorem with experimental data;
- Deepen the understanding of linear circuit superposition, homogeneity and circuit reference direction.

2. The experiment equipment

In this experiment, you will verify Kirchhoff's Laws. The experiment equipment includes DC voltage supplies, multimeter, resistor, slide rheostat and wires.

3. The methods or laws involved in this experiment

Kirchhoff's laws is one of the most basic laws in circuit theory. There are two laws of Kirchhoff: one is the current law and the other is the voltage law.

(1) Kirchhoff's Current Law (KCL): At any time, at any node in the circuit, the algebraic sum of all currents entering and leaving a node must equal zero.

$$\sum_{n=1}^{N} i_n(t) = 0 \quad (6.2)$$

$i_A + i_B + (-i_C) + (-i_D) = 0$

Note that the reference direction of the current is critical for KCL.

(2) Kirchhoff's Voltage Law (KVL): At any time, in any loop in the circuit, the algebraic sum of all voltages is zero.

$$\sum_{n=1}^{N} v_n(t) = 0 \tag{6.3}$$

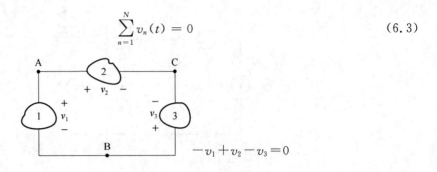

$-v_1 + v_2 - v_3 = 0$

4. Precautions

(1) Note the polarity of the ammeter and the voltmeter.

(2) Select the range reasonably, do not use the meter over the range.

(3) The output of the supply should be increased from small to large.

5. Kirchhoff's Voltage Law

Measure the potential difference between each point orderly in Fig 6.5 and fill in the measurement result in Table 6.5.

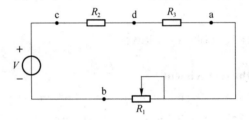

Fig. 6.5 A simple circuit to illustrate KVT

Table 6.5 Measurement for the circuit of Fig. 6.5

Voltage		V_{cd}	R_2	V_{da}	R_3	V_{ca}	V_{ab}	R_1	V_{cb}	V
R_1=_____	Measured Value									
V=_____	Calculated Value	/		/				/		
R_1=_____	Measured Value									
V=_____	Calculated Value	/		/				/		

Result Analysis:

(1) What is the relation among the measured value V_{cd}, V_{da}, V_{ab} and V?

(2) Try to analyze the error, $\left(\text{error} = \dfrac{\text{Measured Value} - \text{Calculated Value}}{\text{Calculated Value}}\right)$

(3) Conclusion:

6. Kirchhoff's Current Law

In Fig. 6.6, V is the independent voltage source, R_1, R_2, R_3, R_4, R_5 are the fixed resistance.

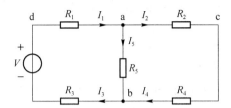

Fig. 6.6 A simple circuit to illustrate KCL

Table 6.6 Measurement for the circuit of Fig. 6.6

Current	I_1	I_2	I_3	I_4	I_5
Measured Value					
Calculated Value					

Result Analysis:

(1) What is the relation among the measured values of I_1, I_2, I_3?

(2) Can you find another set of currents to verify KCL in Fig. 6.6? Prove it.

(3) Try to analyze the error.

(4) Conclusion:

6.3.3 RC circuit and RL circuit

1. The objectives of this experiment

- Learn to use digital signal generator and oscilloscope;
- Understand the RC circuit and the integral circuit.

2. The experiment equipment

In this experiment, you will learn to perform measurements of RC and RL Circuit. The experiment equipment includes DC voltage supplies, digital signal generator, oscilloscope, resistor, capacitor, inductor, LED and wires.

3. The methods, principles or laws involved in this experiment

(1) RC circuit

RC circuits refer to the first order circuits composed of resistance elements and capacitor elements. The simplest RC circuit is a capacitor and a resistor in series (see Fig. 6.7). When the circuit consists of only one charged capacitor and one resistor, the capacitor releases its stored energy through the resistor.

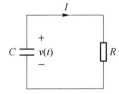

Fig. 6.7 A simple RC circuit

· 201 ·

The voltage across the capacitor, which is time dependent, can be found by using Kirchhoff's current law. This results in the linear differential equation:

$$C\frac{dv(t)}{dt} + \frac{v(t)}{R} = 0 \tag{6.4}$$

where C is the capacitance of capacitor.

Solving this equation for $v(t)$ yields the formula for exponential decay:

$$v(t) = V_0 e^{-\frac{t}{RC}} \tag{6.5}$$

where V_0 is the capacitor voltage at time $t=0$.

(2) *RL* circuit

RL circuits refer to the first order circuits composed of resistance elements and inductance elements. It consists of a resistor and an inductor in series or in parallel. The simplest *RL* circuit consists of an inductor and a resistor in series (see Fig. 6.8).

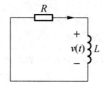

Fig. 6.8 A simple *RL* circuit

The voltage across the capacitor can be found by using Kirchhoff's current law,

$$\frac{1}{L}\frac{dv(t)}{dt} + \frac{v(t)}{R} = 0 \tag{6.6}$$

Then

$$v(t) = V_0 e^{-\frac{tR}{L}} \tag{6.7}$$

where V_0 is the capacitor voltage at time $t=0$.

(3) Integral circuit

An integral circuit is a circuit that makes the output signal proportional to the time integral value of the input signal. The simplest integral circuit consists of a resistor and a capacitor, as shown in Fig. 6.9. The current I is expressed by

$$I = \frac{V_{in}}{R + \frac{1}{j\omega C}} \tag{6.8}$$

Then the voltage of the capacitor V_C is given by

$$V_C = \frac{1}{C}\int_0^t I \, dt \tag{6.9}$$

If the output across the capacitor at high frequency,

$$\omega \gg \frac{1}{RC} \tag{6.10}$$

so

$$I \approx \frac{V_{in}}{R} \tag{6.11}$$

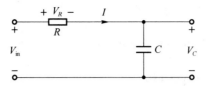

Fig. 6.9　A simple *RC* integral circuit

Now,
$$V_C \approx \frac{1}{RC}\int_0^t V_{in}\,dt \tag{6.12}$$

4. Observe the LED in *RC* and *RL* circuit respectively

(1) Observe the LED in *RC* circuit

Connect the circuit according to Fig. 6.10, where $R = 200\ \Omega$ and $V = 10$ V, then remove the DC voltage supplies and observe the LED, record the phenomenon.

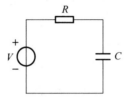

Fig. 6.10　A simple *RC* circuit

(2) Observe the LED in *RL* circuit

Connect the circuit according to Fig. 6.11, where $R = 200\ \Omega$ and $V = 10$ V, then remove the DC voltage supplies and observe the LED, record the phenomenon.

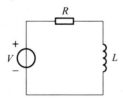

Fig. 6.11　A simple *RL* circuit

5. The input and output of *RC* integral circuit

Connect the circuit according to Fig. 6.12, where the input V_{in} is a sinusoidal signal $V_{in} = V_m \cos(\omega t)$ where $V_m = 5$ V. Observe the output V_C and plot waveform of V_{in} and V_C.

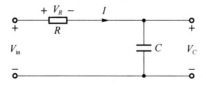

Fig. 6.12　A simple *RC* integral circuit

Plot the waveform of the output.

6.3.4 Verification for Thevenin theorem

1. The objectives of this experiment
- Master the Thevenin equivalent circuit;
- Verify the Thevenin theorem;
- Learn to measure the equivalent parameters of network.

2. The experiment equipment

In this experiment, you will learn to perform measurements of equivalent parameters of network and verify the Thevenin theorem. The experiment equipment includes DC voltage supplies, multimeter, resistor and wires.

3. The methods, principles or laws involved in this experiment

(1) Thevenin equivalent circuit and circuit parameters

Thevenin's theorem states: Any linear one-port network can be equivalent to a resistance in series with an independent voltage source, where the voltage value V_s of the voltage source is equal to the open circuit voltage V_{oc} when terminal a and b are open, and the resistance R_0 of the resistor is equal to the equivalent resistance when all independent sources in the network are set to zero.

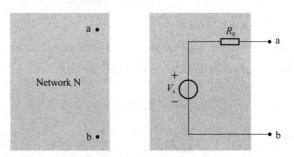

Fig. 6.13 Thevenin equivalent circuit

(2) Measuring method for equivalent resistance of network

① Measuring R_0 by voltametry

As shown in Fig. 6.14, the multimeter is used to measure the voltage at both ends of the resistance R_L and the current passing through it. The external characteristic curve of the network is obtained, as shown in Fig. 6.15. Calculate the slope tg φ according to the external characteristic curve, then the internal resistance can be calculated by

$$R_0 = \text{tg } \varphi = \frac{\Delta V}{\Delta I} = \frac{V_{oc}}{I_{sc}} \tag{6.13}$$

If the internal resistance value of the network is very low, it is not suitable to measure its short circuit current.

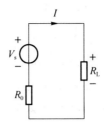

Fig. 6.14 A simple series circuit

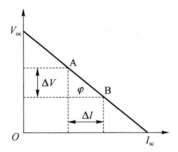

Fig. 6.15 The VCR curve of Thevenin equivalent circuit

② Semi-Voltage method

As shown in Fig. 6.16, when the load voltage is half of the open circuit voltage V_{oc}, disconnect the resistance in the circuit and measure its resistance value with a multimeter. At this point, the value of resistance R_L is the value of resistance R_0 of the equivalent circuit.

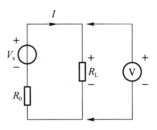

Fig. 6.16 A sinple series circuit

(3) Conditions for the maximum power absorbed by a load

Fig. 6.14 can be seen as a model where a power source delivers electrical energy to a load, R_0 can be viewed as the sum of the source resistance, R_L is the load at both ends of the one-port network, and the power P consumed on the load R_L can be expressed by

$$P = I^2 R_L = \left(\frac{V_s}{R_0 + R_L}\right)^2 R_L \tag{6.14}$$

When $R_L = 0$ or $R_L = \infty$, the power delivered from the source to the load is zero.

Applying different R_L values to Eg(6.14), different P values can be obtained, in which there must be a R_L value to maximize the load's power.

According to the mathematical method of evaluating the maximum value, when $R_L =$

R_0, the maximum power that the load obtains from the source is

$$P_{\max}=\left(\frac{V_s}{R_0+R_L}\right)^2 R_L=\left(\frac{V_s}{2R_L}\right)^2 R_L=\frac{V_s^2}{4R_L} \tag{6.15}$$

At this time, the circuit is said to be in a "match" state.

4. Measurement of Thevenin equivalent circuit via voltammetry

(1) The experimental circuit is shown in Fig. 6.17;

(2) Adjust the DC power supply, to make the output voltage $V_s=12$ V;

(3) Measure the open circuit voltage V_{oc} at both ends of the network;

(4) Measure short-circuit current I_{sc};

(5) Connect the load R_L, change the load R_L according to the requirements in the table, and measure V_{ab} and I;

(6) Using the data in the table, draw the volt-ampere characteristic curve $V_{ab}=f(I)$, and then use voltammetry to calculate R_0;

(7) Obtain Thevenin equivalent circuit of network N according to the open circuit voltage and load R_0.

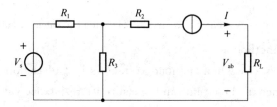

Fig. 6.17 A simple circuit to measure Thevenin equivalent circuit

Table 6.7 Tabulation of V_{ab} and I at different value of R_L for the circuit of Fig. 6.17

R_L/Ω	100	200	270	470	600	1 000	open circuit
V_{ab}/V							
I/mA							

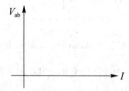

From the curve, we can obtain $R_0=$ _____ Ω, $V_{oc}=$ _____ V.

5. Measurement of Norton equivalent circuit via Semi-Voltage method

(1) The experimental circuit is shown in Fig. 6.18;

(2) Measure the short circuit current I_{sc} of the circuit shown above;

(3) Measure the open circuit voltage V_{oc} of the circuit shown above;

(4) According to the Semi-Voltage method, adjust the sliding rheostat; when the voltage is half of V_s, disconnect the R_L from the circuit and measure it the resistance value

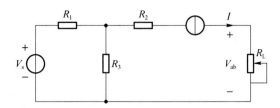

Fig. 6.18 A simple circuit to measure Norton equivalent circuit

with a multimeter. The resistance value of R_L is the resistance of the equivalent circuit R_0;

(5) Change the sliding rheostat R_L, and measure the corresponding V_{ab} and I' for different resistance values. Record the data and plot the voltammetric characteristic curve;

(6) Get Norton equivalent circuit of the network N according to the short-circuit current I_{sc} and R_0.

Table 6.8 Measurement for the circuit of Fig. 6.18

R_L/Ω	100	200	270	470	600	1 000	open circuit
V_{ab}/V							
I/mA							

$R_0 =$ _____ Ω, $I =$ _____ mA.